The Memory Bible

The Memory Bible

An Innovative Strategy for Keeping Your Brain Young

Gary Small, M.D.
Director of the UCLA Center on Aging

HYPERION New York

ISBN 0-7868-6826-0

Designed by Lorelle Graffeo

Hyperion books are available for special promotions and premiums. For details contact Hyperion Special Markets, 77 W. 66th Street, 11th Floor, New York, New York 10023-6298, or call 212-456-0100.

FIRST EDITION

10 9 8 7 6 5 4 3 2 1

Contents

Preface

Nearly everyone struggles with some form of memory loss before reaching middle age. Thanks to recently developed brain-imaging and genetic technologies, scientists can now observe the earliest physical indicators of brain aging in people as young as twenty-five. Tiny plaques and tangles that develop and grow ever denser in our brains often begin accumulating decades before any middle-age forgetfulness sets in. A minute spot of plaque on a 30-year-old brain could possibly indicate Alzheimer's disease forty years from now, just as a tiny little snag of the dentist's probe can mean a cavity in the making.

But we need not despair. Misplacing your keys a couple of times doesn't mean you should start labeling your cabinets. Memory loss is not an inevitable consequence of aging. Our brains can fight back, and *The Memory Bible* will give you the tools. We can improve our memory performance immediately and stave off, possibly even prevent, future memory decline. The sooner all of us begin our memory program, the sooner we will be on the path to keeping our brains young and healthy for the rest of our lives.

Gary Small, M.D.
Los Angeles, California, May 2002

Acknowledgments

I am grateful to many people who helped make this book a reality: my wife, best friend, and collaborator, Gigi Vorgan, who helped translate a scientific treatise into this book; my children, Rachel and Harry, who make me laugh every day and keep my heart young; and my parents, Gertrude and Dr. Max Small, who always see me as young, no matter how beyond middle-aged I become. Several friends and colleagues provided input and encouragement, including Helen Berman, Susan Bowerman, Dr. Josh Chodosh, Dr. Linda Ercoli, Stuart Grant, Dr. David Heber, Diana Jacobs, Dr. Lissy Jarvik, Dr. Jim Joseph, Richard Hissong, Andrea Kaplan, Dr. Michael Phelps, Dr. Stephen Read, Dr. John Schwartz, Dottie Sefton, Don Seigel, and Pauline Spaulding. I am particularly grateful to my editor, Mary Ellen O'Neill, who kept me focused on what is important, and my long-time agent and friend, Sandra Dijkstra.

NOTE:

The stories contained in this book are composite accounts based on the experiences of many individuals and do not represent any one person or group of people. Similarities to any one person or persons are coincidental and unintentional. Readers should consider consulting with their physician before initiating any exercise or treatment program.

The Memory Bible

Chapter One

You Have More Control Than You Think

I have a photographic memory but once in a while I forget to take off the lens cap.

—Milton Berle

Imagine struggling your way out the glass doors of a crowded mall in late December, loaded with shopping bags, packages, and presents. Your head is pounding and your feet hate you *and* the shoes you walked in on. You'd die of starvation this second if you weren't already dying of thirst. You manage to pull out your car keys and glance up at the humongous, jam-packed parking structure when it hits you—you've forgotten where you parked.

Could never happen to you, you say? Ever forgotten your purse, wallet, file, or phone at home, only to remember it while caught in rush hour traffic? Maybe you've struggled to remember the name of a movie you saw last night or that new neighbor you just met, not five minutes ago. Ring a bell?

Most of us laugh off these so-called middle-aged pauses, considering them just another normal annoyance of aging, not a true memory problem, and certainly not a sign of Alzheimer's disease—not at "our age." I hate to pop another "I could party all night and still get to work on time if I wanted to" baby-boomer bubble, but it's time for us all to wake up—we are *all* one day closer to Alzheimer's disease.

It Is Never Too Late or Too Early to Fight Brain Aging

Just as all of us inevitably get older, recent convincing scientific evidence shows that Alzheimer's disease is not simply an illness that some old people get. Alzheimer's disease or a related dementia may well be *everybody's* end result of brain aging—and it begins forming in our brains much earlier than anyone previously imagined, even in our twenties.

The subtle, gradual aging of the brain starts as tiny plaques and tangles that begin accumulating there, decades before a doctor can recognize any symptoms of the disease. In fact, these plaques and tangles begin forming so early in our adult lives that subtle memory and language changes go unnoticed and ignored for many years. Nonetheless, these minuscule spots of plaque in our otherwise healthy brains are the first signs of brain aging, and they will increase insidiously if we do nothing about them.

When I speak on this subject I am often asked: Will my brain already be irreversibly damaged by the time I reach middle age? Is it too late for me to try to head off this inevitable process? Is it too soon for me to start? Is my memory ability destined to decline no matter what I try to do about it?

My answer to these questions is no. It is *never* too late or too early to start beating the brain-aging game. Even if one day research finds a way to restore already lost brain cells, scientists agree that *preventing* the loss of memory will always be easier than restoring it. The sooner we rise to the challenge, the sooner we can intervene in the battle, like little neuron-gladiators, and, with luck, do so while our forgetfulness is minimal or even imperceptible.

Our Brains Aren't Getting Any Younger, but They Can Get Better

One of the biggest obstacles to starting a program to improve memory performance and protect our brains from Alzheimer's disease is denial that one's brain, as well as one's body, is aging. Many people struggle to accept the physical changes that come with passing years, yet coming to terms with mental changes is often an even greater challenge.

Sally B. had a reputation as a fabulous hostess—her parties had always been the talk of the town. For several weeks her daughter had been reminding her to prepare a guest list for her sixty-fifth birthday party, but Sally just kept forgetting to do it. Finally her husband Jerry mentioned that she was forgetting quite a few things lately and suggested she discuss this with their family internist.

Sally scoffed at Jerry's "accusation" and told him that *her* mind was *perfectly fine*. If anyone was getting old and losing their marbles, it was *he*. It was just that the thought of a sixty-fifth birthday didn't seem right for someone like her. She didn't *feel* 65, and thanks to Dr. Mark, she sure didn't look it. In fact, lots of people at the club said that she and her daughter looked just like sisters.

In the last ten years, Sally had undergone two face-lifts, cheek and chin implants, liposuction, breast lift and augmentation, innumerable Botox and collagen injections, and a tummy tuck. She was a regular at Dr. Mark's surgery center and had met almost every anesthesiologist, nurse, and orderly that worked there.

Jerry, still going on about her birthday, insisted Sally allow them to throw her a wonderful party for a change—*she* wouldn't have to do a thing! Sally laughed. "Of course I will. I'll have to call Dr. Mark and have my eyes done right away."

Jerry hit the roof. "No more calling Dr. Mark, Sally. You can't have plastic surgery every six months. It's not good for you."

Sally looked hurt and responded indignantly, "I haven't had anything done in two years."

Jerry said softly, "Honey, what about the tummy tuck five months ago? Don't you remember? You couldn't walk for two weeks."

Sally thought about it. "Oh, yeah, right. Well, that had nothing to do with my *face*. Besides, that Linda Bens . . . Dens . . . something, at the club, gets a face-lift every year for Christmas, and she looks just fine."

Jerry cut a deal with her. If she'd accompany him to their family doctor to discuss her memory changes, he'd back off about the surgery. Sally agreed.

The internist performed a standard memory test on Sally in the office and was concerned enough to send her to a geriatric psychiatrist to get a more detailed assessment. After several meetings, the psychiatrist sat down with Sally and Jerry together. Sally was indeed suffering from some mild age-related cognitive impairment. As difficult as that was for her to accept, thankfully there were plenty of things she could do to fight it; however, getting another face-lift or eye job surely wasn't going to help.

The psychiatrist recommended she start taking a cholinergic medication, begin a memory-training program, and try various other strategies to protect her brain. He explained to Sally that success with this treatment required acceptance and a strong commitment. The psychiatrist also expressed concern about unnecessary elective surgeries because repeated general anesthesia can potentially worsen memory ability.

Sally began walking a half hour every day and taking medication to improve her memory. She started a program for keeping her brain young that included antioxidant foods and stimulating mental aerobics. In several weeks, Sally, as well as her family and friends, noticed improvement in her memory and her mood.

Sally had a wonderful time at her sixty-fifth birthday party, as did everybody else. And she and her daughter looked just like sisters.

Beginning a program to improve memory and slow down brain aging requires accepting that we need such a program. A better understanding of what actually happens to our memory abilities and our brains as we age will help us keep our brains at their peak performance.

What Is Memory?

Normal memory performance involves both learning and recall (Figure 1.1) and requires intact functioning of several regions of the brain and the brain cells, or neurons, within them. We generally think of memory as an abstract concept—a thought, image, sensation, or feeling that is stored somewhere in our brain's filing cabinet, ready to be pulled out at will. However, because our brains are comprised of nerve cells, chemicals, and electrical impulses, our memories are actually encoded, stored, and retrieved as a result of minuscule chemical and electrical interactions.

Each nerve cell in the brain has a single axon that acts like a telephone line, conducting nerve impulses toward neighboring neurons.

The friendly neuron next door receives the countless assortment of electrical impulses sent to it daily, through its dendrites—bunches of thin filaments extending out like little antennae, receiving and sending information. But the new info is not home free yet.

To allow all of our brain's neurons to communicate with the others, the axons and dendrites form thousands of branches, and each branch ends in a synapse, a specialized contact point or receptor that recognizes only extremely specific information being passed between neurons. Each neuron has approximately 100,000 synapses.

Electrical nerve impulses containing the new information, retrieved memories, or relayed messages shoot down the neuron's axon and slip through one of its skinny dendrites into a hyper-specific synapse, where a packet of chemicals, known as neurotransmitters, gets released. These neurotransmitters are the "carrier pigeons" that travel the minuscule space from one synapse to the next. Upon arrival, the correct chemical neurotransmitter binds with its corresponding receiver, and voilà! The message is received. In this way, thoughts and ideas are conveyed, information is learned, and memories are retrieved, all of which cause us to do, think, or act in different ways.

In any waking situation, our senses are bombarded by sights, sounds, and other stimuli that pass through our *immediate memory* and move into a holding area known as *short-term memory*. We usually lose most of these fleeting sensations in milliseconds, and of the few retained in our short-term memory, only a small percentage ever make it into long-term memory storage.

An essential key to retaining information longer is to organize and rehearse it, thus actively working it into our *long-term memory*. Some people require great effort to develop these skills, while others are born with a knack for memory techniques and "tricks" to reinforce new information and make it stick. They are often considered to have "photographic memories"—a myth we shall discuss later.

Once information is lodged in our long-term memory, it

becomes relatively permanent and can be recalled years later—as long as our brains remain healthy. While short-term memory has only limited capacity, long-term memory has the potential to store tremendous amounts of information. Retrieving this information later, or pulling it out of memory storage, is known as *recall*. Even patients with advanced Alzheimer's disease, who may have difficulty remembering their morning meal, have been known to recall long-ago events, such as their first date with a sweetheart, in vivid detail.

Figure 1.1

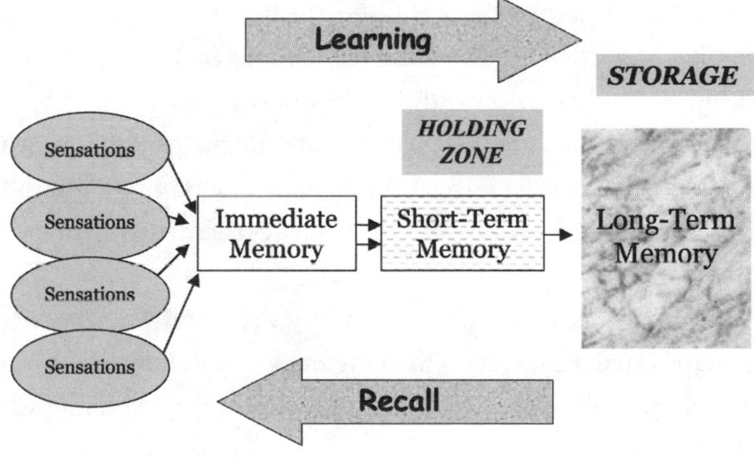

Recently, scientists have learned how the brain converts short-term memories into permanent ones at the molecular and cellular level. A specific protein must be present in the brain's cerebral cortex, the outer rim of the brain containing gray matter, for the process to succeed.

The brain's hippocampus, a seahorse-shaped brain structure located in the temporal lobe of the brain (near the temples), stores information on a temporary basis—much like a computer holds data in its random access memory. When the brain converts the information into permanent memory, similar to writing data to a computer's hard disk, the hippocampus interacts with the cerebral cortex to complete the task.

Sex, Style, and Emotion

People vary in their learning styles. Long before I became interested in memory research, I instinctively relied upon my visual learning strengths in everyday life. I had always found it easier to remember someone's last name if I spelled it out in my mind's eye. Auditory learners retain information best if they hear it, while visual learners remember best when they actually see the information.

Memory and other cognitive skills often vary according to gender: women tend to have better verbal and language abilities, while men generally have the edge in spatial and mathematical abilities. However, when I mentioned this to my wife, she nearly managed to talk me out of it.

Various other factors influence our memory abilities. Emotional states have a major impact on the efficiency and the quality of memories. Ask yourself where you were and what you were doing when President Kennedy was shot. (Or John Lennon, for you youngsters.) All of us who were around certainly know the details of where we were, whom we were with, and how we felt, yet I doubt that we can remember similar details of events the week before. Information that is emotionally charged has a distinct quality and is easier to learn and recall. The memory of your first crush in second grade probably remains distinct. Many of us can recall details of that boy or girl we barely knew decades ago. By contrast, when we are experiencing feelings of depression and prolonged anxiety or stress, we become distracted and our memory abilities diminish.

Memory Changes with Age

Although we all experience some forgetfulness as we age, we each differ in our degree of memory change, our concern about it, and the steps we take to cope. By the time we reach our thirties and forties, so-called "normal" memory complaints become more common.

Middle-aged and older people most often notice difficulties with:

- **People's names**
- **Important dates**
- **Location of household objects**
- **Recent and past events**
- **Meetings and appointments**
- **Recalling information**

Age-related memory loss more often involves recent memories rather than distant, past ones. We might forget what movie we saw last weekend yet still recall our ninth-grade homeroom teacher's name. Neuropsychological evidence shows that age tends to slow down our learning and recall skills, perhaps making it more difficult for older adults to learn a foreign language or scientific discourse. (I wouldn't want to try to pass advanced calculus again at 50.)

Older people have greater difficulty multitasking and our reaction time can slow down as we age, which can affect our daily activities. Many older drivers compensate by driving more slowly, which can be a hazard in itself. Memory training (Chapters 3 and 6) and a program of mental aerobics (Chapter 5) can help lessen the impact of many of these age-related changes.

In the early 1990s, memory experts defined diagnostic criteria for the memory changes that accompany normal aging. When someone over 50 had a memory impairment demonstrated by at least one standard memory test, along with a subjective awareness of memory changes, they called the phenomenon *age-associated memory impairment*. These experts estimated that 40 percent of all people are affected by this condition upon reaching their fifties, 50 percent in their sixties, and over 70 percent by age 70 and older.

Although there is debate over whether or not age-associated memory impairment will or will not progress and at what rate, it is likely that the condition precedes other, more severe memory declines.

Without some form of intervention, whether it's implementing strategies to keep your brain young and healthy or, if needed, medical evaluation and treatment, people who ignore their age-associated memory impairment may eventually develop *mild cognitive impairment*. An estimated 10 million Americans over age 65 suffer from this more severe memory decline, and this condition has a 10 to 15 percent chance of developing into Alzheimer's disease with each year that passes.

Figure 1.2

PROGRESSION OF AGE-RELATED MEMORY LOSS

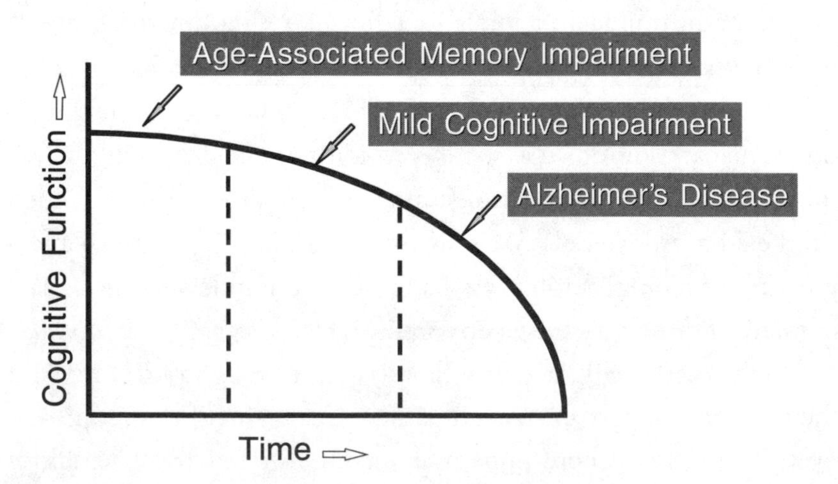

These diagnostic categories—age-associated memory impairment, mild cognitive impairment, and Alzheimer's disease—are basically categories of convenience, allowing doctors and scientists to better understand our aging brains and test treatments to alleviate memory decline (Figure 1.2). In reality, the changes in our brains and the memory difficulties we experience are continuous, fluid processes beginning remarkably early in our lives. Several recent studies point to just how early.

Language Skills and IQ Tests at 20 Predict Alzheimer's Disease at 80

Several years ago, Dr. David Snowdon and his associates at the University of Kentucky performed clinical evaluations and standard memory tests on a group of nuns who were aged 70 or older. Each of these participants in what is known as *The Nun Study* kept diaries when they entered their convents in their early twenties. The scientists had access to these earlier documents and performed a standardized linguistic analysis of these diaries, objectively rating early language ability. The nuns whose youthful writings demonstrated greater idea density and grammatical complexity were much less likely to develop significant memory loss or Alzheimer's disease decades later, in their seventies.

The study's conclusion, that language ability at age 20 may predict whether or not someone will get Alzheimer's disease fifty years down the road, stirred debate over whether learning and educational enrichment protected the brain from decline over time, the "use it or lose it" theory.

More recently, Scottish psychiatrist Dr. L. J. Whalley and his colleagues studied intelligence test records to determine if a person's IQ early in life predicted Alzheimer's disease up to fifty years later. This group found that people with lower intelligence test scores in childhood had a greater risk for the late-onset form of Alzheimer's disease that begins after age 65.

Dr. Whalley offered several explanations for the observation, including the possibility that people with lower intelligence in childhood might engage in behaviors later in life that put them in greater danger of getting Alzheimer's disease. They may eat a less healthy diet, avoid exercise, or smoke. Alternatively, the low IQ score may reflect the early signs of the disease itself deteriorating the brain subtly early in life. This could then influence school performance and further educational pursuit. Having less education may not be the cause but may actually be the result of the early stages of brain aging.

The Incredible Shrinking Brain: Beware of Plaques and Tangles

As our brains age, the synapses, or connections between neurons, begin to function less efficiently. Messages firing from one region of the brain to another may get scrambled, and crucial communication from one part of the brain to the other may break down. One area of your brain may tell you to walk into the kitchen and open the refrigerator, but then you just stand there. Unfortunately, the part of the brain that should have told you to reach in and get a soda because you're thirsty didn't receive the message.

Data show that as our neurons age and die, the actual overall sizes of our brains shrink or atrophy. Also, our aging brains accumulate lesions known as *amyloid plaques* and *neurofibrillary tangles*. These collections of decayed material result from cell death and degeneration of brain tissue, particularly in areas involved in memory: the temporal (under the temples), parietal (above and behind the temples), and frontal (near the forehead) regions of the cerebral cortex, the outer layer of brain cells. A healthy, plump brain containing only sparse plaques and tangles gradually shrinks to an atrophied Alzheimer's brain riddled with plaques and tangles.

Historically, a definitive diagnosis of Alzheimer's disease could only be made at autopsy. The pathologist would count up the number of plaques and tangles that had accumulated in these key brain regions, and if their concentration surpassed the defined threshold the patient under examination definitely had Alzheimer's disease. Scientists have studied brain autopsies in people who had only mild cognitive impairment rather than Alzheimer's. They see the same plaques and tangles, in the same brain areas, only in lower concentrations.

These autopsy studies have now been extended to people in their twenties and thirties who had normal memory abilities, and still these brain lesions are seen to be present, albeit in lower concentrations. In every age group, the accumulation pattern is consistent: the lesions start in areas near the temporal lobe and spread to the parietal and frontal regions. Most of us, unless of course we have a genetic risk or some other predisposition, don't live long enough to reach the plaque-and-tangle threshold defined as full-blown Alzheimer's disease.

Studies of the annual incidence of Alzheimer's disease, or the percentage of the population that develops it each year, show that the rate of new Alzheimer's cases doubles every five years between ages 65 and 90. Scientists suspect that if the current trend toward increased lifespan continues, people may soon be living, on average, well into their eighties and nineties. Unfortunately, the proportion of the population with Alzheimer's or another dementia will rise correspondingly. In fact, I am convinced that, if we did nothing to prevent brain aging, the prevalence of Alzheimer's disease would approach 100 percent if we all lived to be age 110 (Figure 1.3).

Figure 1.3

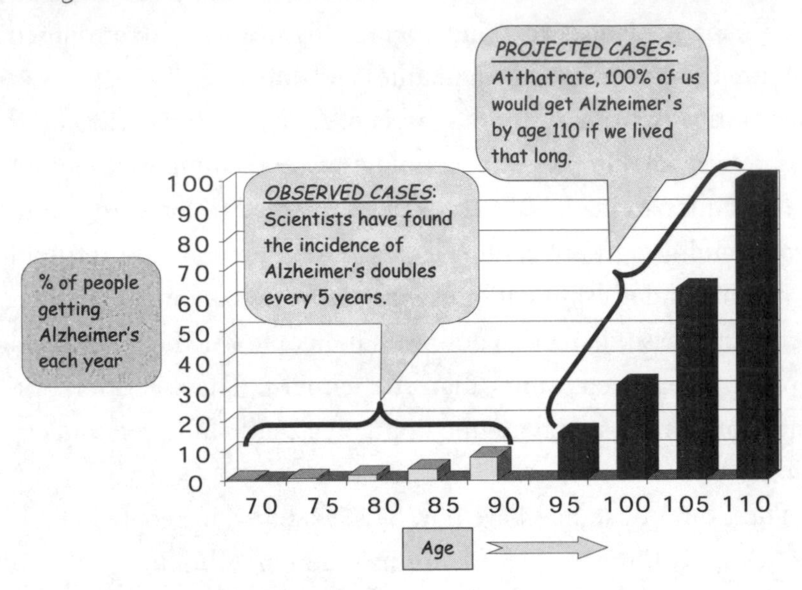

The rate at which our brains age varies according to our individual genetic predisposition, lifestyle choices, and our lifelong environmental exposures. Also, the use of new technological advances allows us to recognize the earliest signs of brain aging without having to dig up our old high school diaries or agree to a brain biopsy.

Big Heads Don't Make Men Smarter

Subtle and not so subtle differences between women and men likely influence memory abilities and brain health as they age. Women have smaller brains than men. (It's just a fact, don't shoot the messenger.) The average brain weight for an adult man is just over three pounds, while the typical woman's brain is a bit over two and two-thirds pounds. Neuroscientists have found that generally the bigger the brain, the smarter the animal, but that rule does seem to break down with the human brain—a point my better half will argue adamantly.

Recent studies of brain structure and function have shown that although women have smaller brains, their brains are more efficient, thus leveling the overall intellectual abilities between women and men.

Dr. Ruben Gur and his colleagues at the University of Pennsylvania looked at the amount of gray matter in the brain—the outer part containing cell bodies that allow us to think—and found that, on average, 55 percent of a woman's brain contains gray matter, compared with only 50 percent of a man's brain. This may explain why women score higher in language and verbal ability tests than men. By contrast, men have a higher proportion of white matter, which transfers information from distant regions, perhaps a key to their greater visual-spatial abilities.

Granny's Not Sick, She's Just Old and Getting Senile

When I was growing up, Billy J., the kid across the street, had his grandmother living with them. Every once in a while, she would wander out of the house, and Billy's parents would have to go looking for her with the car. One time they didn't find her for an entire day. When my father, a physician, asked Billy's dad if he could recommend a doctor to help her, Billy's dad laughed and said, "Granny's not sick, she's just old and senile."

Early in my clinical and research training in geriatric psychiatry and Alzheimer's disease, Dr. Lissy Jarvik stressed that senility was not a normal part of aging but instead a *disease*. This was an important message at the time because most experts were ignoring the problems of aging, even the most common ones: memory loss and dementia. By emphasizing the disease factor, investigators began to approach the problem as an abnormality that required accurate diagnosis and specific treatment.

In fact, this has long been the basic approach of western medicine, to diagnose a disease and look for the best treatment and cure.

It is still a challenge to initiate proactive, preventive approaches to diseases. We all want a quick cure, a magic pill to alleviate our problems when they occur. Patients *and* physicians are reluctant to "fix it if it ain't broke." But when it comes to an aging brain, what we don't know *will* hurt us.

Understanding senility to be a disease state, whether in its early or late forms of dementia or as full-blown Alzheimer's disease, was crucial to getting researchers to focus on how our brains change with age and the problems that can and do arise. With today's knowledge, as well as new tools that allow scientists to *see* the brain changing at its very earliest stages, the future lies in research and treatment to help slow or halt these changes and some day repair any existing damage.

Of course, we must all first face our own fears about what our memory problems may imply and any stigma we attach to "mild forgetfulness." With the understanding that brain aging is a human phenomenon that affects us all, hopefully people will begin using *The Memory Bible*'s strategies to become proactive about preventing memory loss and protecting their brains from Alzheimer's disease. Those with more pronounced memory loss conditions may become empowered to come forward and begin using new brain-imaging technologies for early detection and treatment of dementias and Alzheimer's disease.

Memory Training—Brain Fitness

If you're reading this, you are most likely seeking knowledge on how to maintain a young and healthy mind, maximize your memory performance, and protect your brain from Alzheimer's disease. The memory program described in the chapters ahead will help you accomplish these goals. What's more, you will see your memory improve as soon as you get started.

Nearly a decade ago, neuroscientists studied brain scans of volunteers playing the computer game Tetris for the first time. They found high levels of brain activity. A month later, when the volunteers had become proficient at the game, their scans displayed significantly lower levels of brain activity. This lower brain activity, indicating greater mental efficiency, tells us that with time, practice, and familiarity, our brains essentially adapt themselves to achieve the same results with less work. The process is similar to what occurs when people train their muscles by lifting weights—their bodies eventually develop muscular efficiency. Bench-pressing the same barbell will require much less effort after a month of training, and most athletes have to add more weight if they wish to continue strengthening.

Knowing that our brains can become more efficient if we practice or become skilled at memory techniques, we can begin to systematically *train our brains*. By using games, puzzles, and some new approaches to daily activities, we can improve our short- and long-term memory abilities and possibly prevent future memory loss and Alzheimer's disease.

Studies have shown that memory training, an integral part of any program to slow brain aging, benefits more than just mild forgetfulness and overall memory. The training also gives the user an awareness of their improvement, allowing them to *feel good* about their enhanced learning and recall skills, which in turn improves their memory performance even more.

Recent research points to various forms of mental activities, vocational occupations, and educational achievement as a means to decrease our risk for future memory decline and eventual development of Alzheimer's disease. People engaged in mentally challenging jobs or pursuits are somehow more protected from future memory losses.

Scientific cause and effect has been proven in the laboratory using mature animals, half of which were allowed to live in a mentally stimulating and exciting environment with mazes, toys, and hidden surprises and snacks. The other half was exposed to dull, standard-issue laboratory living environments. Although brain size in these mature laboratory animals generally shrinks with age, the animals exposed to mentally stimulating environments had higher numbers of neurons in the memory areas of the brain as well as better learning abilities than the experimental animals in the less interesting settings. If these findings hold true for humans, they point to continued mental activity throughout life as a strong preventative for future cognitive decline.

Joe T., an insurance agent, and his wife, Alice, a school administrator, used their savings to buy a town house overlooking the fifth hole of a beautiful golf course in a desert retirement community, six hours away. For almost five years they used it on weekends and holidays, usually inviting friends or family along to golf and enjoy the views. Joe was ecstatic when he and Alice finally qualified for early retirement—at last they could stop working, move to the town house, and start living the good life while they were young enough to enjoy it.

Alice cried at her office farewell party—after twenty-six years, her co-workers had become like family and she had enjoyed the daily challenge of coming to work and "putting out fires." As they finished packing, Joe was more excited than a kid on his first trip to Disneyland. He kissed Alice, the love of his life, as he carefully folded an array of new Hawaiian shirts and swore he'd never wear a suit and tie again.

The first six months passed quickly as Joe relandscaped, installed an outdoor grill, and perfected his golf swing. Alice got busy redecorating the town house and entertaining their frequent houseguests. Before the year was out, Alice, who had already cut back on golf due to the heat, was sidelined altogether by a sprained ankle. They got cable TV so she could watch her favorite old movies.

Their daughters and grandchildren visited less frequently now, as did their friends. Alice understood that everybody had hectic work and school schedules and that nobody could get away to the "good life" every weekend, but she still got lonely. Joe encouraged Alice to get involved with the country club and other local groups.

Alice tried, but she was never fond of playing bridge and soon grew weary of the country club's events and the community's women's organizations. She spent hours on the phone with friends and family and checked in with her old office at least once a week. She missed her old life and was just plain bored.

She began having trouble sleeping through the night and needed to nap during the day. Alice became withdrawn and depressed, and Joe couldn't understand it. He tried to make her see the bright side—they were healthy, they were in love, and they were living their dream. Alice wondered if she had ever actually had this dream. Perhaps it was just that Joe had wanted it so badly that she started wanting it too, because she loved him. Well, it hardly mattered now.

Alice's memory lapses started slowly—forgetting a barbecue at the clubhouse, mixing up the arrival dates of visitors—but they were soon noticeable to Joe and their daughters. Alice had never before missed a birthday call to the grandkids or forgotten to buy half the items she needed at the market. She was worried and told Joe that either she was going senile or the desert heat was cooking her brain.

They consulted a local doctor to help figure out why these memory problems had come on so quickly. After all, she had gotten rid of the stress and responsibilities of her job and had less she needed to remember now than she used to. After examining Alice, the doctor concluded that her memory lapses might be due to boredom and general lack of mental stimulation. It sounded to him like maybe what she needed was to get a job. Joe immediately said that was ridiculous, they were retired now, but Alice was intrigued by the idea.

She was reluctant to make Joe alter his dream in any small way, but in marriage one learns to compromise, and they had always done it well.

There was a job opening for an executive administrator at the local school board, and Alice easily landed it. Her memory improved, her ankle healed, and for her the good life included the daily mental challenge that came with work and productivity.

For Alice and many like her, mental stimulation is crucial to mental health and memory performance. She had thrived on a certain level of mental stimulation, and once that was removed, she swiftly declined. Memory training is actually a focused form of mental stimulation that allows you to efficiently pack a big memory punch in a short amount of time. Even if memory training doesn't ultimately prevent Alzheimer's disease, it will improve current memory ability. It is eminently achievable. One of the greatest benefits of memory training is that it gives us tools to use throughout our lives. If we master the techniques early on, we have a better chance of heading off memory loss that might emerge in the future.

Windows to Your Brain: New Technologies to Detect Brain Aging

Despite our best efforts at gyms and beauty salons, the physical results of aging are fairly obvious: wrinkling skin, graying hair, even disappearing hair for many of us. In contrast, brain aging is a much greater challenge to detect. Scientists have searched for decades to find a way to view brain structure and function, so as to pinpoint a problem that might improve with treatment and determine the specific treatment, when to intervene, and whether the patient was benefiting from it.

Recently there have been research breakthroughs from diverse scientific disciplines—genetics, chemistry, physics, biomathematics, as well as others—that are finally opening windows into the brain, using new technologies like positron emission tomography, or PET, scanning. We can now view brain aging directly and thereby specifically guide our treatments to prevent future memory loss.

During medical school in the mid-1970s, I remember our excitement the first time we viewed computed tomography, or CT, scans of the brain. We finally had a way to look at brain tissue beyond what conventional X-ray machines could provide. With the development of magnetic resonance imaging, or MRI, we could see even more detailed brain images—enabling us to diagnose strokes, tumors, and hemorrhages. Although these innovative techniques provided information on brain structure or shrinkage, indicating that brain cells had already died, they offered no information about how well the still-living brain cells were functioning. If only we could actually see how effectively the neurons were or were not communicating with each other, then we might be able to pick up, and treat, more subtle brain deficiencies before the cells died. Thanks to my UCLA colleague Dr. Michael Phelps and others, we now have the stunning breakthrough discovery of PET scanning and can finally see this kind of subtle brain dysfunction in living humans.

Positron emission tomography reveals a consistent pattern in Alzheimer's disease. The parietal and temporal areas—where Alzheimer's first strikes—show reduced activity in the early stage of the disease. It looks as if those important brain memory centers are subtly and gradually fading away (Appendix 2). The PET scan is currently the most sensitive technology for making an accurate, early diagnosis for guiding treatment. In our UCLA Memory Clinic, we use it to diagnose Alzheimer's disease years before most doctors would be able to confirm the diagnosis with conventional methods.

There are many of us who suffer from much milder memory symptoms, and my UCLA research team wondered whether these new technologies could help us to recognize more subtle brain aging. By combining PET scanning and information on a person's genetic risk for Alzheimer's disease, we uncovered a way to observe very mild brain aging—the changes that are occurring today in many baby boomers. These tools may also help us to gauge the success of our memory fitness program and other interventions at slowing that brain-aging process down.

DNA before AARP*—Genetics of Brain Aging

The science of genetics has ballooned in the last fifteen years. Most of us know that genes are the blueprints of life, and everybody's DNA differs just enough to make us all individuals—more reliable, even, than fingerprints.

When we think of genetic traits being passed from one generation to the next, we usually think of physical features such as hair and eye color, facial features, height and build, and so on. It is only in recent years that medical conditions such as heart disease, high cholesterol, and cancer have been discovered to pass within families genetically.

**Formerly known as American Association of Retired Persons.*

Traditionally, the common, late-onset form of Alzheimer's disease, which affects people after age 65, was not thought to have a genetic influence, but to be a normal result of aging. We now believe the cause involves a combination of environmental, lifestyle, and genetic influences.

Many genes have been discovered to be involved with age-related memory loss and Alzheimer's disease. A defect in some genes causes early-onset familial Alzheimer's, a rare and devastating form of the disease that hits people early in life, before age 65, and normally strikes half the relatives in those families.

For the common late-onset Alzheimer's, however, one major genetic risk has been discovered: apolipoprotein E, or APOE. This APOE gene makes a protein that transports cholesterol and fats through the body and is known to influence the risk for heart disease and related conditions, so it was a big surprise when geneticists found a link to Alzheimer's disease and memory loss. APOE comes in three different forms, or alleles: APOE-2, APOE-3, and APOE-4. All of us inherit one APOE form or allele from each parent for a combination of two alleles, known as a genotype.

Drs. Allen Roses, Margaret Pericak-Vance, and their co-workers at Duke University were the first to show that the APOE-4 allele was much more frequent in Alzheimer's patients than in normal people. Approximately 65 percent of the population has the APOE 3/3 genotype; 20 percent has the 3/4 genotype—a high risk for developing late-onset Alzheimer's disease; and 2 percent the 4/4 genotype—an even higher risk for Alzheimer's disease.

Although the APOE-4 gene increases a person's risk for Alzheimer's and makes it more likely that they'll get the disease at a younger age, an APOE blood test result alone is not enough to accurately predict whether an individual will get the disease. The research team I direct at UCLA recently achieved a significant advancement in understanding the brain-imaging window using

PET by linking it to the latest research in genetics. Combining these scientific technologies for the first time provided a key for early detection of subtle brain changes related to aging, changes that may precede the onset of Alzheimer's disease by several decades.

Prevention Is the Best Medicine

By the time a patient develops Alzheimer's disease, the damage is done and irreversible. In the absence of a "cure," our best shot at beating Alzheimer's lies in prevention, and targeting mild forgetfulness is where we at UCLA started. A large challenge has been to determine which of the millions of forgetful people would be most likely to benefit from preventative treatments.

In the mid-1990s, our research group at UCLA combined information about brain function and genetics in our studies of people with only mild memory complaints. With aid from the National Institutes of Health and in collaboration with the Duke University genetics group, we found that middle-aged people *without* Alzheimer's disease still displayed decreased brain function in the parietal region of the brain—an area important to memory—*if* they possessed the APOE-4 genetic risk. Dr. Eric Reiman and his collaborators at the University of Arizona independently confirmed our discovery. More recently, we discovered that this decreased brain activity worsens over the years. Everyone with the APOE-4 genetic risk for Alzheimer's disease showed decline in brain function on follow-up PET scans in the parietal and temporal memory areas of the brain.

These discoveries have provided us a standard way to detect decline in brain function long before people get to end-stage brain aging, that is, Alzheimer's disease. In response to these breakthroughs, moreover, we have created the UCLA Memory Clinic, an innovative clinical and research program wherein we test new treatments to see if we can slow down the decline in brain function in adults at any age, from 20 to 100 years.

For a preventative treatment to be considered effective, it must slow down the rate that brain activity declines over the years compared with a placebo or an inactive intervention. In Figure 1.4 below, the solid line indicates the rapid decline of someone receiving a placebo, while the dotted line shows the slower decline of someone receiving an active drug that prevents brain aging as measured by a PET scan. Using this approach, we are testing several drugs in middle-aged and older people, and we are beginning similar studies to test the effects of memory training and mental aerobics on stabilizing brain function and slowing brain aging.

Figure 1.4

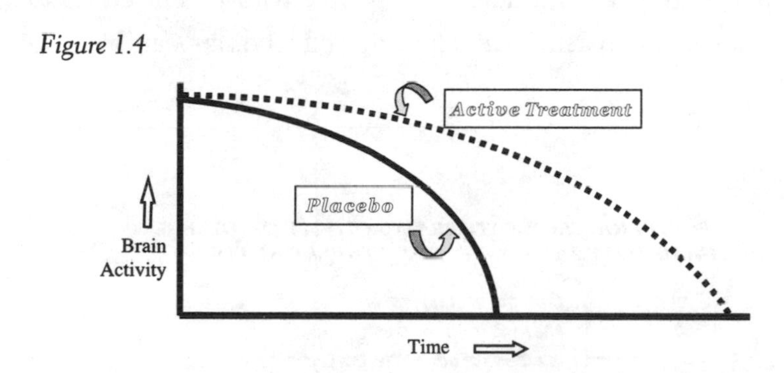

College Grads Get Higher (. . . Brain Activity, That Is)

Following up on the conclusions from *The Nun Study*—that subtle differences in the language abilities of young people might predict the development of Alzheimer's disease fifty years later—my research team attempted to use PET scans to detect such brain function deficits in young adults. Dr. Daniel Silverman and I looked at the scans of people with normal memory abilities and compared their brain function in the posterior cingulate memory center, according to whether or not they had finished college. We found that those who had completed college displayed higher brain function during mental rest; however, this increased brain activity diminished with age. The 50-year-old college grads in our study had much higher activity levels than the 50-year-olds who did not complete college, whereas the 80-year-old college grads had only a very slight, if any, increase above 80-year-olds who did not complete college. Age had worn away their brain function reserve.

Based on these findings, Dr. Silverman and I systematically reviewed the PET scan results of a much larger group of young adults according to their educational achievement. We now have results of PET scans showing the brain activity of people in their twenties—the same age as the nuns in David Snowdon's study when they displayed differences in language ability predicting Alzheimer's disease fifty years later. Figure 1.5 illustrates how people with higher education have enhanced brain activity and how this effect disappears with age.

Figure 1.5

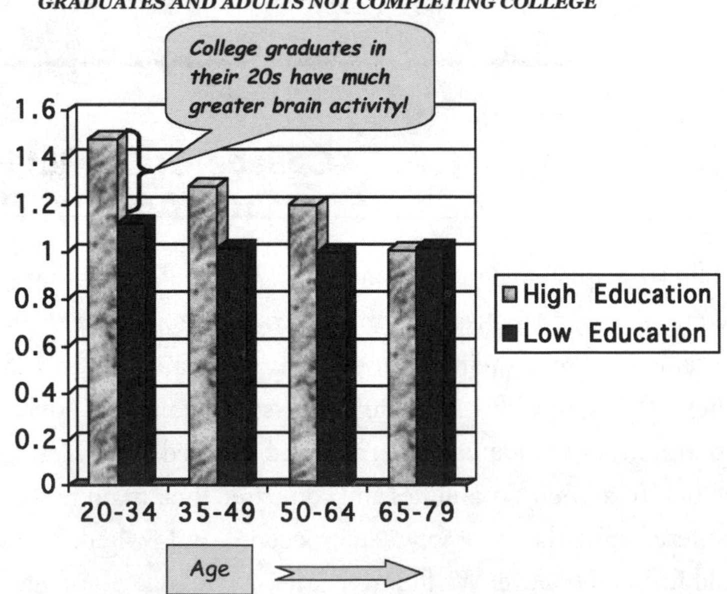

Our study not only demonstrated subtle patterns of brain functional reserve in young adults, but it supported the idea that subtle brain changes can be observed in people beginning in their twenties, an age when Alzheimer's disease wouldn't normally strike for another forty or fifty years. When we looked at the effect of higher education added to the effect of the Alzheimer's APOE-4 genetic risk, we found, as expected, that the young adults with the *greatest* brain activity had *completed* college and did *not* have the APOE-4 gene. It is interesting to note, however, that the influence of a college education on a person's brain activity reserve was even more powerful than that of the APOE-4 risk gene. Ideally, someone concerned about their APOE-4 genetic risk could compensate for possible brain activity deficits through further education.

These observations are consistent with autopsy studies of young individuals aged 22 to 46 years that have found the earliest stages of collections of plaques and tangles in 36 percent of those with the Alzheimer's APOE-4 genetic risk, compared with only 11 percent of those without the genetic risk.

As yet, there is no conclusive evidence on the mechanism by which education protects brain cells. A recent study of brain autopsies found that people with lower educational attainment had more evidence of vascular disease in their brains compared to those who had attended college. Educated people may be less inclined to smoke, drink too much, and eat fatty diets, which could keep their brains healthier. Our research group's discovery is also consistent with the "use it or lose it" theory—it is also possible that having a healthier brain to begin with sets one on the college trajectory.

The Cognitive Stress Test

My research group at UCLA also has developed a cognitive stress test, similar to a cardiac stress test that a cardiologist might give a heart patient, whereby the patient exercises on a treadmill to stress the heart, in order to bring out subtle cardiac abnormalities not observed on an electrocardiogram performed during rest. In devising the cognitive stress test, we asked volunteers to perform memory tasks during brain scans to observe subtle brain alterations not observed by the scanner during mental rest. Dr. Susan Bookheimer and I used functional MRI scans, which show brain activity during mental tasks.

About half of our study's volunteers had the APOE-4 genetic risk for Alzheimer's disease, while the other half did not. Everyone performed the same memory tasks during their scans, which included learning and then attempting to recall a series of unrelated word pairs (e.g., author–tree, table–elephant). All the participants performed this task relatively well, but those with the Alzheimer's APOE-4 genetic risk required a significantly greater amount of brain activity to accomplish the same memory task. The areas of the brain that worked the hardest to complete the memory tasks were the very same areas where Alzheimer's disease initially strikes. In fact, within the hippocampus, one of the brain's main memory centers, brain activity for the at-risk volunteers was double that of people without the genetic risk. Not surprisingly, our cognitive stress test was fairly accurate in predicting which participants eventually developed further memory loss several years later.

PET Scanning and Genetic Testing: Are They Ready for Prime Time?

Many experts anticipate that Medicare and private insurers will soon make reimbursement available for PET scanning in the diagnosis of Alzheimer's disease and other memory problems. This would be a great step forward for preventative care because PET technology detects the signature pattern of Alzheimer's disease in people without severe memory impairment, thus guiding appropriate early treatment. Anyone with a new concern about increased forgetfulness or a sudden change in memory ability should consult a physician and, if indicated, get a PET scan. You can find a local PET center through the Academy of Molecular Imaging (Appendix 5).

Most doctors do not recommend APOE genetic testing for only mild memory complaints. If someone receives a clinical diagnosis of Alzheimer's disease, then an APOE test might be recommended to increase diagnostic accuracy. In those rare families where nearly half of relatives develop Alzheimer's disease before age 65, more extensive genetic counseling and testing is generally recommended.

We are continuing our work on combining PET scanning and genetic risk testing to assess new therapies for mild memory complaints that would prevent further brain function decline and memory loss. At the same time, geneticists are getting closer to identifying additional genetic risks for Alzheimer's disease as we continue to refine and automate our biological brain-imaging technologies.

At our UCLA Memory Clinic we have seen encouraging results in people using our memory-training and mental aerobics exercises, which incorporate games, puzzles, and other mentally stimulating activities. It is critical that each of us begins memory training at our own level. The key is to find yours and get started. Even with a tight schedule and limited time, you will benefit from memory training—if only for a few minutes a day. And as you build skills, the benefits will keep getting greater.

Chapter Two
Rate Your Current Memory

When I was younger, I could remember anything, whether it happened or not.

—Mark Twain

To effectively begin a program to improve memory performance and keep our brains young, we first need to rate our current memory ability level. Knowing where our memory abilities stand now directs us as to where to begin our overall anti–brain-aging program, enabling us to set an initial, easily attainable goal to ease us into the program.

Elliot S. was an accomplished 68-year-old statistician—a member of the National Academy of Sciences and among the short list of mathematicians in line for the Nobel Prize. He spent the first thirty years of his career working in a prestigious think tank before accepting an endowed chair from an ivy league university. It was hard for family and friends to notice his memory decline since it was so gradual and he was such an introvert. Elliot's records showed that in 1972, his IQ tested at 160—among the top 1 percent. At the time of his first visit to our memory clinic in March 2001, his IQ had declined to 115. Although much lower than his peak performance, he was still among the top 5 percent of the population. Therefore, someone meeting Elliott for the first time would never guess what his PET scan revealed to us: he had advanced Alzheimer's disease.

The system by which we assess our current state of brain aging can vary, from simple self-assessment questions to more detailed biological measures. Although a PET scan is the most sensitive technique to uncover early brain aging, I am not recommending that every one of us run out and get one tomorrow. For most people, this chapter's memory rating system will suffice. It will also help individuals to find the correct level at which to begin their personal anti–brain-aging program. If upon completing this assessment, you feel you need additional consultation, see Appendix 5 to find organizations that provide local and national resources and referrals.

Subjective Versus Objective Memory Loss

Objective memory is how well we actually perform on a pencil-and-paper memory test. *Subjective* memory is our own perception of how well *we think* we do in memory functions. Both types of memory assessment are important in understanding the type of memory changes each individual is experiencing as well as in setting up a personalized memory fitness program.

Audrey M. watched her grandfather and aunt succumb to Alzheimer's disease before their sixtieth birthdays.

Although both her own parents died young in a car accident, she was convinced that she too would develop the disease at a young age and in her early forties began seeking a specialist. Audrey insisted on consulting the busiest neurologist in town, and hounded him to complete an extensive test battery she had read about on the Internet. Even though the doctor saw no indication for such an evaluation, he went along with her demands and ordered an EEG, MRI, APOE screening, and other laboratory tests, and a neuropsychological workup.

The doctor left for Europe after explaining that he'd give Audrey feedback when he returned and had the test results in hand. His first day back at the office, he checked his voice mail and found several messages from Audrey:

Message 1: Hello, doctor, this is Audrey M. You told me you'd be back in the office Thursday at 3 p.m. It's now Friday, 10 a.m., you are not there, and neither are your nurses, Ilene, Carol, or Wendy. Please call me.

Message 2: It's Audrey M., and I misplaced my keys again, doctor. I'm getting much worse, my memory is going. You must have all my test results by now. The APOE test was supposed to take longest, and that was due back last Wednesday by 4 p.m. Please, please call me.

Message 3: This is Audrey. I just lost my address book. One minute it was here; then it disappears into thin air. I don't remember touching it. Thank god I know your phone number by heart. Call me the moment you get in.

The doctor checked Audrey's test results, which showed no signs of Alzheimer's disease. When he informed her, she broke down in tears of relief.

Clearly, Audrey's anxiety and worry had colored her subjective memory awareness. Because of her family history, she was convinced that every minor memory lapse was a

drastic onset of her worst fears. The doctor pointed out to her that even without her neuropsychological test results, her voice mail messages demonstrated a remarkable attention to detail and high level of objective memory ability.

Each individual is more or less aware of memory changes over a lifetime. Memory experts have found many factors that influence how seriously we take these changes and whether or not we will complain about them to others. A person's mood and sense of well-being influence how much they notice and complain about forgetfulness. Depression and anxiety also increase self-awareness of memory difficulties. College graduates and others with higher education tend to rate their memory abilities better than do those who have never attended college. Of course, people are more likely to complain about forgetfulness as they get older. At a certain age, conversations, jokes, and complaints about middle-aged pauses and senior moments are a socially correct form of bonding, just like young parents' complaints about their toddlers or teenagers' complaints about their parents.

Memory scientists have developed standardized questionnaires for determining the degree to which each of us is aware of memory loss. Because so many different factors can influence awareness of memory loss, some question whether these so-called *subjective* memory measures accurately reflect true memory changes or whether they merely reflect each person's distortions, prejudices, moods, and concerns rather than their actual memory ability.

My research group has performed extensive studies of these self-awareness measures and found that they indeed reflect an objective, biological change. In one of our studies, we used a particular memory self-rating questionnaire developed by Dr. Michael Gilewski and his associates at Cedars-Sinai Medical Center in Los Angeles. We asked a group of middle-aged and older adults with only mild memory complaints to complete this questionnaire. We found that people with a greater subjective awareness of memory loss had a much greater likelihood of also possessing the gene associated with Alzheimer's disease risk—the APOE-4 gene. And both this genetic risk and increased awareness of memory loss predicted future decline in objective memory ability.

More recently, Dr. Daniel Silverman and I looked at a group of same-aged people with mild memory complaints who had received a PET scan when we evaluated how they rated their own memories. Two years later, we repeated their PET scans and found that those volunteers who *believed* their recall abilities were worsening actually exhibited significantly decreased activity in the hippocampus memory center of the brain on the later scan.

These findings emphasize the "go with your gut" approach to detection of brain aging. If you think your memory loss is "all in your head," it may well be true, and worth looking into.

Rate Your Memory Change Awareness

To discover how aware you are of your own memory changes, you need a measuring system. Here we will use a modified version of the self-rating questionnaire we have used in our research studies at UCLA.

Answer the questions in the following Subjective Memory Questionnaire by circling a number between 1 and 7 that best reflects how you judge your own memory ability. Afterward, we will tally your results to use in determining the best training level for you to begin your memory program.

SUBJECTIVE MEMORY QUESTIONNAIRE

How would you rate your overall memory?	**Poor**		**Good**			**Excellent**	
	1	2	3	4	5	6	7
How often do these present a problem for you?	**Always**		**Sometimes**			**Never**	
names	1	2	3	4	5	6	7
faces	1	2	3	4	5	6	7
appointments	1	2	3	4	5	6	7
where I put things (e.g., keys, eyeglasses)	1	2	3	4	5	6	7
performing household chores	1	2	3	4	5	6	7
directions to places	1	2	3	4	5	6	7
phone numbers I have just checked	1	2	3	4	5	6	7
phone numbers I use frequently	1	2	3	4	5	6	7
things people tell me	1	2	3	4	5	6	7
keeping up correspondence	1	2	3	4	5	6	7
personal dates (e.g., birthdays)	1	2	3	4	5	6	7
words	1	2	3	4	5	6	7
forgetting what I wanted to buy at the store	1	2	3	4	5	6	7
taking a test	1	2	3	4	5	6	7
beginning something and forgetting what I was doing	1	2	3	4	5	6	7
losing my thread of thought in conversation	1	2	3	4	5	6	7
losing my thread of thought in public speaking	1	2	3	4	5	6	7
knowing whether I have already told someone something	1	2	3	4	5	6	7

As you read a novel, how often do you have trouble remembering what you have read?	**Always**		**Sometimes**			**Never**	
in opening chapters, once I've finished the book	1	2	3	4	5	6	7
three or four chapters before the one I'm now reading	1	2	3	4	5	6	7
chapter before the one I'm now reading	1	2	3	4	5	6	7
paragraph just before the one I'm now reading	1	2	3	4	5	6	7
sentence just before the one I'm now reading	1	2	3	4	5	6	7

How well do you remember things that occurred . . .	**Poorly**		**Fair**			**Well**	
last month	1	2	3	4	5	6	7
between six months and one year ago	1	2	3	4	5	6	7
between one and five years ago	1	2	3	4	5	6	7
between six and ten years ago	1	2	3	4	5	6	7

When you read a newspaper or magazine article, how often do you have trouble remembering what you have read?	**Always**		**Sometimes**			**Never**	
in the opening paragraphs, once I have finished the article	1	2	3	4	5	6	7
three or four paragraphs before the one I am currently reading	1	2	3	4	5	6	7
the paragraph before the one I am currently reading	1	2	3	4	5	6	7
three or four sentences before the one I am currently reading	1	2	3	4	5	6	7
the sentence before the one I am currently reading	1	2	3	4	5	6	7

Add up all the numbers you have circled. If your total score is 200 or more, then your subjective memory difficulties are minimal. You may find that you quickly master the Three Basic Memory Training Skills (Chapter 3) and can move right on to more advanced memory skills training (Chapter 6). If your score is between 100 and 200, then you are noticing a moderate degree of memory challenge. You may want to spend more time on developing basic memory skills (Chapter 3) before moving on to advanced memory skills training. A total score below 100 reflects an even greater self-awareness of memory difficulties. A score in this range would suggest that memory training will be a greater challenge, so it is important to take your time with the exercises in the following chapters. You also might consider contacting your physician about these concerns or one of the experts in your area (Appendix 5).

Objective Memory Ability

An objective memory test assesses our current learning and recall abilities. The traditional, extensive types of objective memory assessments—neuropsychological testing—can take hours to complete and require highly trained professionals to administer, score, and interpret. I developed the following simple, do-it-yourself objective memory-rating method that you can complete right now. This brief version of the more extensive assessment we use in our research and clinical work emphasizes retrieval or recall of words you will learn during the test. And, recall—the ability to pull that information out of your memory storage—is *the* major area of concern for most people. This objective assessment measure will complement the results of your subjective memory assessment and give you a clear idea of where to begin, as well as focus, your program.

Do not be discouraged if you find this memory assessment

method too difficult or perhaps too easy. It is designed to assess memory in people with a wide range of abilities. It is also intended to be difficult at first to enable you to see concrete results from your memory skills training program soon after you start it. I guarantee that your score will improve immediately after reading Chapter 3.

The Objective Memory Test

Because the assessment is timed, you will need a stopwatch or kitchen timer or timepiece with a second hand before beginning. The test involves learning a list of ten words over a 1-minute period and recalling them after a 20-minute break. When ready, set your timer for 1 minute, then read and learn the words on the list in Assessment No. 1.

Assessment No. 1
Study the following words for up to 1 minute:

Plank
Banker
Sauce
Umbrella
Abdomen
Reptile
Lobster
Orchestra
Forehead
Jury

When your minute is up, put aside *The Memory Bible*, reset your timer for a 20-minute break, and do something else—read a newspaper, start a crossword puzzle, whatever you like, just make sure you distract yourself from the word list with something else. After the 20 minutes, write down as many of the words as you can recall.

To interpret your score, add up the number of correct words you can write down after 20 minutes of distraction. If you did well on your objective memory recall score (8 or greater), then the basic memory skills (Chapter 3) will probably be quite easy for you to master and you can quickly move on to advanced memory skill training (Chapter 6). If your score is less than 8, then you need to spend more time on learning basic memory skills before moving on. If your score is lower, below 4, don't panic. This assessment tool is designed to be difficult for many people. Move on to Chapter 3 and then see how you do on your retest. If your objective memory recall score improves, then continue to build your memory skills program. If not, you might consider contacting your physician or an expert in your area for a professional evaluation (Appendix 5).

Many factors can influence your objective memory score, particularly your age and level of educational achievement. In general, younger people score better than older people, and people who have had more educational experience will have better scores. The results are an indicator, a guide, not the last word on your current brain fitness.

Ironically, a common cause of memory complaints is worry and anxiety about memory performance. People with a family history of Alzheimer's disease may fret about every memory slip. In fact, worry about memory difficulties may worsen actual objective memory performance. If you scored well on objective word recall after 20 minutes but your subjective memory score indicated frequent memory difficulties, you may be suffering from stress and anxiety. If you fit into this group, then I suggest turning to Chapter 4 (Minimize Stress) before beginning basic memory training skills (Chapter 3).

Table 2.1 summarizes the information you need to decide at what level to begin your program.

Table 2.1

INTERPRETING SUBJECTIVE AND OBJECTIVE MEMORY SCORES

MEMORY SCORE		
Subjective	Objective	Program
High	High	Move quickly from basic (Chapter 3) to advanced (Chapter 6) memory training.
Low	Low	Take time and focus on basic training, then reassess score. If no improvement, consult expert.
Low	High	Focus on stress reduction (Chapter 4) before basic memory skills training. Reassess score and consult expert if no improvement.

When you have determined your initial, or baseline, subjective and objective memory performance scores, plot them in the charts in Figure 2.1. After completing the basic memory training (Chapter 3), you will be asked to take the objective test again. To assess your progress, return to this figure, enter your score, and congratulate yourself on your improved score after learning only a few basic skills. Try repeating both the subjective and objective tests again after completing Chapter 10. I believe you will see a clear and steady rise in both your subjective and objective performance as you continue to learn and practice the memory training skills and initiate your program of mental aerobics, stress reduction, and more advanced memory training.

Figure 2.1

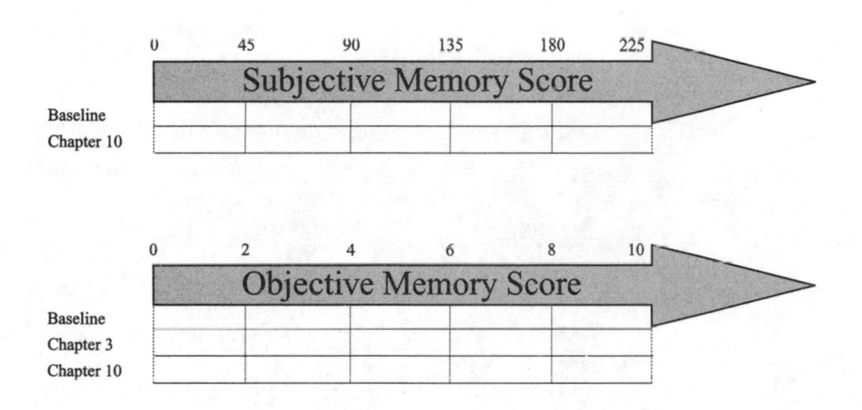

Chapter Three

LOOK, SNAP, CONNECT:

THE THREE BASIC MEMORY TRAINING SKILLS

Everything should be made as simple as possible but not simpler.
—ALBERT EINSTEIN

Most of us have seen people with so-called photographic memories—they can rapidly learn and recall the order of cards from a shuffled deck or effortlessly memorize long lists of words or numbers. However, there is no such thing as a truly photographic memory; what we have seen are people with good *memory techniques*.

A memory technique is merely a coding system, a filing cabinet for the brain. A person's prior knowledge and interests will influence how well they learn and recall new information. Many teenagers can readily recall scores of their favorite sports teams but cannot remember a single important date they learned in history class last fall. (You know who you are.)

Our memory's filing and storage systems work most effectively when the new information contains meaning for us. Experiments with champion chess players show that they can readily memorize the chess pieces on the board if they are placed as they would be during a match, but the players' ability to recall a pattern of pieces placed at random is almost impossible. One arrangement of chess pieces has meaning while the other does not.

Such meaning may actually be hard-wired in our brains. When people focus on meaningful words as opposed to strings of random letters, functional MRI brain scans show increased activity in specific areas of the frontal and temporal lobes. The greater the activity, the greater the likelihood that they would remember the words later. Great memories are not born, they are made.

By mastering my three basic skills—*LOOK, SNAP, CONNECT*—you will incorporate a foundation for a solid memory-training program. If you read only this chapter and learn only these three skills, your memory will improve. I suggest you keep a notebook or writing pad specifically dedicated to your memory-training exercises. This way you will be able to see your progress and keep track of your exercises.

1. *LOOK*—Actively Observe What You Want to Learn

I had a friend in college who was not only a brilliant mathematician but also a gifted writer and violinist. He was at the top of the curve in all our pre-med classes. With his many talents and intellectual abilities, he still had extreme difficulty remembering people's names and connecting the names to the faces when he met them again. Eventually he realized that his problem was that he never really learned the names in the first place—he wasn't actively looking and listening.

One of the most common barriers to effective learning is that people do not pay attention to situations in which new information is presented. Think about what your husband or wife wore to work this morning. Can you remember what tie he had on? Which blouse she wore? What color your son's T-shirt was? By actively looking and making a conscious effort to take in this type of information, trivial as it may seem, you can begin to train your brain to log in details. By engaging in this active observation process, we can absorb details and meaning from a new face, event, or conversation, which helps us to learn and remember it.

When we have little interest in something, it is often difficult to remember because we are not paying full attention. Many of us forget names seconds after being introduced to someone new. It's as if we're on automatic pilot, responding to various internal and external cues during the introduction, which distract us from retaining any new information. It is essential to slow down—just enough—to notice what is being said and whether it is important to remember. Samuel Johnson put it succinctly: "The true art of memory is the art of attention."

Memory for street directions is an excellent example. If you drive yourself to a new destination by following directions, you'll probably remember how to get there on your own, days or even weeks later. If you were merely a passenger on your first trip, you are likely to get lost on your first solo trip. The goal in active observation (see Active Observation Exercises box) is to mentally stay in the driver's seat.

LOOK is the first basic skill because our vision is so often our first exposure to the things we want to remember, although we rely heavily on our other senses as well. By repeating information for later memory retrieval, we are listening. Many report that the sense of smell can bring back the most vivid memories of all. Textures and temperatures are useful details for focusing our observational skills that employ sense of touch for future recall. *LOOK*, as it is used in the Three Basic Skills, is actually shorthand for all five senses: *LOOK*, *LISTEN*, *FEEL*, *TASTE*, and *SMELL*.

Active Observation Exercises

1. The next time someone drives you to an unfamiliar location, mentally put yourself in the driver's seat. Check the directions ahead of time, note the street signs, major intersections, and landmarks. Mentally drive yourself back there later.
2. Before you see a new movie, make a conscious decision to remember certain details at the outset. Try to commit to memory the hero's hairstyle, the furniture in a memorable indoor scene, and the full name of a supporting character. When you get home, jot down as many details about the movie as you can recall. The next day, check your list and try to write down even more details about the movie if you can.
3. At work, notice a detail about the clothing or general appearance of several co-workers. Write down the person's name in one column and the detail in another. At the end of the day, cover the second column, look at each name, and try to remember the specific details.
4. Right now, try to remember one specific article of clothing each member of your family put on this morning before leaving the house. Tomorrow do it again, but make a conscious effort to observe details before leaving the house.

2. SNAP—Create Mental Snapshots of Memories

Go back to that image of what your mate wore to work this morning. Red blouse, black slacks and shoes, leather jacket. As you visualize the image, you are already developing the second basic skill—*SNAP*.

You are creating a *mental snapshot* of the information you wish to remember. Afterward you just pull out the snapshot and describe what you see. Creating vivid and memorable images fixes them into our long-term memory storage.

Snaps can take two forms, *real* or *imagined*. A real snap involves active observation, concentrating on what you actually see and making a conscious effort to fix the observed image into a mental snapshot. *Imagined* snaps are those that you create from your own memories and fantasies, but they still become fixed in your memory as a mental snapshot. *Imagined* snaps can be a fantasy distortion of an image you observe.

We all have real snaps in our heads that we use instinctively to help us seek and retrieve lost or hidden objects. Our ability to use search images effectively likely evolved as an adaptive advantage in our hunting ancestors, who visually spotted their skirting prey with such images. Likewise we use them to quickly retrieve that partially hidden green folder on the desk or the slightly soiled lucky basketball jersey in the hamper. Despite the disorganization and ridiculous overcrowding of my home library shelves, I may have little trouble spotting a particular tome if it has a strong search image fixed in my memory—"it's a thick blue book with two white lines and some circle-like flower things on it." In the bookshelf below, the search image of one particular book helps it to stand out.

Children are naturals in using their imagination. They have active fantasy lives that tend to diminish with age. As adults, we are taught to suppress this natural ability in exchange for more controlled, logical thinking. A child's whimsical fantasies and imagery might be considered psychotic thinking in the average adult. To help develop effective learning and recall techniques, we need to rekindle these natural creative instincts.

Bright, colorful, enhanced snapshots stick best in memory, as do those with movement, three dimensions, and detail. The more detailed the image, the easier it will be to recall later. The very act of focusing on detail helps us to pay better attention and learn the information contained in the image. Compare the two drawings of sandwiches below. The more detailed version will be easier to remember.

Distorting or exaggerating one or more aspects of your snaps can also give them personal meaning, making them easier to learn and to recall later. The more vividly and creatively we visualize new information for ourselves, the more effectively it will stick in our minds. Imagination can be as outrageous, vivid, or sensual as we like, as long as it enhances our ability to store and recall information.

You need to buy a pumpkin after work for your daughter's Halloween party, so try personalizing it in your mind. Your daughter loves to wear her pearl necklace to parties, so you could picture one draped on a pumpkin, like so:

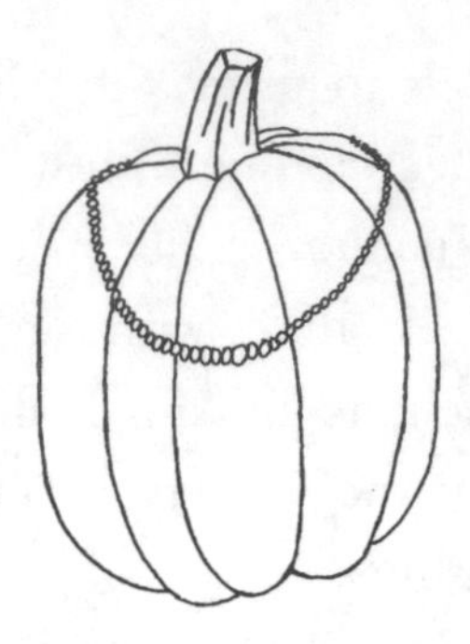

Or you can create a literal image—whereby you write out in your mind the word you wish to remember—but for most people, a symbolic image containing personal or emotional meaning works best. If I park my car on level 3B of a parking structure, I could create a three-dimensional visual image, or mental snap, of 3B as follows:

But for me, a more effective strategy would be to visualize an image of *three bumblebees* hovering over my car. I have a personal aversion to bees, and it would be unpleasant to approach my car with three giant bees hovering above it. I take a mental snap of that image, and the emotional charge of my mental snap helps to fix it in my memory. For someone else, visualizing *three bears* sitting in their car may have a greater emotional impact—perhaps that was their favorite children's story. *Three Buffalo Bills* might work for you—as long as it helps you find the car. See the Mental Snapshot Exercises box for ways to develop effective visualization techniques.

Mental Snapshot Exercises

1. Sit in a comfortable chair, close your eyes and think of the first thing that comes to mind, be it an object, situation, person, or animal. Now try to imagine greater detail about how it looks or feels.

2. For each of the following, create a colorful, vivid, and detailed visual image (e.g., rather than a rose, visualize a bright yellow rose with dew drops, lady bugs, and thorny stems):

    - **Animal**
    - **Kitchen appliance**
    - **Tool**

3. Visualize each of the following but alter them slightly so they become unusual in some way (e.g., an automobile with a wig, a snake wearing suspenders):

    - **Table**
    - **Stadium**

- **Stethoscope**
- **Airplane**

4. List three details that you might see in a visual image of each of the following places:

    - **Shopping mall**
    - **Car wash**
    - **Soccer field**
    - **Church**

3. *CONNECT*—Link Your Mental Snapshots Together

Developing techniques to connect mental snaps together is a basic element of nearly all memory techniques. *CONNECT* is the process of associating two mental Snaps so you can remember the connection later. This basic skill will help you to remember birth dates, the names of employees' spouses, and allow you to never again forget the name connected to the face (Chapter 6).

To connect two snaps, simply create a brand-new snap that contains both mental images. Several techniques can make *CONNECT* an extremely effective memory tool (see box).

Techniques to Effectively Connect Mental Snapshots

- **Place one image on top of the other**
- **Make one image rotate or dance around the other**
- **Have one image crash or penetrate the other**
- **Merge or melt the images together**
- **Wrap one image around the other**

In the example above, we see two ways to connect two mental snaps, one of trees and one of a helicopter. You want to remember that the helicopter is over the trees. The literal image on the left may help, but I doubt you will ever forget the merged, wacky image on the right.

Before proceeding, try a couple of exercises to help fine-tune your own connecting abilities (see box).

Connecting Exercises

1. For each of the following word pairs, imagine a situation or activity that involves the two together. Try to create a situation that is reasonable or logical in some way.

   Telephone—hamper
   Paper clip—stuffed animal
   Apple—policeman
   Stethoscope—football

2. Now go back to the word pairs above and imagine a bizarre or illogical situation for each.

CONNECT is the basis of the *link method*, which orders items by associating the things-to-be-remembered with each other—the ideas or images become part of a chain, starting with the first item, which is associated with the second, the second with the third, and so forth. When initiating the first link in the chain, be sure that item number one helps you recall your goal or reason for creating this particular list.

Linking often helps when we need to remember a list of unrelated things to do, particularly if writing out the list is inconvenient or impossible (you may be exercising at the gym, sitting in a conference, or in the middle of a shower).

If we need to remember a long list of items, the link method becomes a more elaborate method of connecting mental snapshots and becomes a story. The story's flow and visual images provide the cues for retrieving the information. A weakness of the link method is that if we forget one link, we can forget all the information that follows. With the *story system*, the flow of the story will allow most of the remainder of the list to be retrieved even if one link is broken.

Here is a typical list that can be linked into a story line:

- **Buy eggs**
- **Call your cousin in New York City**
- **Take out the garbage**
- **Feed your neighbor's dog**
- **Get cash at the ATM**

To link the ideas, we first choose a single image to represent each task:

- **Egg**
- **A Big Apple**

- **Garbage can**
- **Dog**
- **Dollar bills**

You know you can only begin the errands after work, so that might be the starting point of the linked associations. To help the information stick, you may want to add vivid or emotional detail to the images. The following might be your sequence of linked images:

- **Driving home from work you see a giant egg in the middle of the road**
- **The egg roles down a driveway and smashes against a Big Apple**
- **You pick up the mess and toss it into a garbage can**
- **As you close the garbage can, your neighbor's dog sniffs at it curiously**
- **The dog sits up and faces the can with dollar bills in his mouth**

One of the limitations of linked associations is that the images may not immediately bring to mind the task. For example, you may ask yourself, "What about that big apple? Was I supposed to make a pie? Prove yet again Newton's Law of Gravity? Perhaps make a call to my cousin in New York City?" A better reminder for phoning your cousin in New York might be an image of a big apple talking on the phone.

The most effective links or associations are ones we create ourselves, particularly those stemming from our first association. Psychoanalysts have used the method of free association to help people uncover emotionally charged experiences. Often our first association to an idea is the most vivid and can have the strongest emotional charge or personal meaning, making it easiest to remember.

Another application of *CONNECT* is the use of acronyms, or the creation of words from the first letters of items to be remembered. To create an acronym, first think of one word to represent each item to be remembered, and then form a word using the first letter of each word to be remembered. As an example, for the list of unrelated words **E**nvelope, **B**ulb, **A**ardvark, and **T**elephone, we might use the first letter of each word—*E*, *B*, A, and *T*—to form the acronym. The next step is to write down the letters and play with their order to try to come up with one word or several words. If you can't come up with a word, then try substituting one of the words to be remembered. For example, changing *bulb* to *light bulb* allows us to substitute *L* for *B* and come up with the acronym *LATE*. People who like doing word jumble puzzles will enjoy using the acronym memory method (see the More Connecting Exercises box).

The more you practice *LOOK, SNAP, CONNECT* to help remember tasks, events, and lists of any type, the more familiar and natural it will become. These three fundamental skills are the building blocks for the advanced memory training outlined in Chapter 6 (Build Your Memory Skills Beyond the Basics). But first, check your progress since reading this chapter.

More Connecting Exercises

1. Think up a story that will connect the following items: helicopter, movie theater, library, houseboat, grandmother, coffee mug. After you complete the next five linking exercises, see if you can recall the items from your story, without looking back at this list.
2. Create a single visual image to link each of the following groups of words:

   Wire–teddy bear–rose bush
   Lamp–potato–motorcycle
   Keyboard–cowboy–blimp
   Apple sauce–parachute–hitchhiker

3. List five things you need to do tomorrow.
4. Now use the link system to recall the list above.
5. Now use the story system for the same list.
6. Create a one-word acronym by using a first letter from each of the following items:

   Elephant
   Blanket
   House
   Apple
   Tree

Another Shot at the Objective Memory Test: This Time with Ammo

After reading and practicing *LOOK, SNAP, CONNECT*, your memory abilities have already improved. I have revised the objective memory test from the end of Chapter 2 to include a different list of words so you can retest your learning and recall skills. Get out your stopwatch or kitchen timer and set it for 1 minute, then read and learn the words on the new list in Assessment No. 2.

ASSESSMENT NO. 2
STUDY THE FOLLOWING WORDS FOR UP TO 1 MINUTE:

Ink
Kettle
Spray
Musician
Volcano
Monarch
Steamer
Dirt
Lawn
Gallery

When your minute is up, put aside *The Memory Bible*, reset your timer for a 20-minute break, and distract yourself by doing something else. After 20 minutes, come back and write down as many of the words as you can recall. Compare the number of words you wrote down to the results of your earlier assessment and marvel at your improved memory performance (Chapter 2).

LOOK, SNAP, CONNECT: A Quick Review

1. ***LOOK**—Actively Observe What You Want to Learn.* Slow down, take notice, and focus on what you want to remember. Consciously absorb details and meaning from a new face, event, or conversation.

2. ***SNAP**—Create Mental Snapshots of Memories.* Create a mental snapshot of the visual information you wish to remember. Add details to give the snaps personal meaning and make them easier to learn and recall later.

3. ***CONNECT**—Link Your Mental Snapshots Together.* Associate the images-to-be-remembered in a chain, starting with the first image, which is associated with the second, the second with the third, and so forth. Be sure the first image helps you recall the reason for remembering the chain.

Chapter Four

Minimize Stress

Pressure and stress are the common cold of the psyche.
—Andrew Denton

Our modern world is filled with new technological tools designed to make our lives more efficient and stress-free—computers, voice mail, e-mail, cell phones, hand-held organizers, and tiny digital devices to record our output. For many of us, having all this modern equipment has not only failed to limit the stress in our lives but instead appears to have increased it. Because these technologies allow us to be more efficient, we can now take on more activities and responsibilities, which in turn put even greater demands on our personal and professional lives. Tick tock, tick tock. Time is money. Get more for less.

Chronically high levels of stress are not only bad for blood pressure, cholesterol, and other physical ailments, but such stress levels wear away at brain fitness and overall memory performance. You've had a tough week at work; you're kids are driving you up the wall with their bickering; that new contractor botched up the remodeling of your kitchen; and your formerly compulsive memory for all the details seems to be getting worse by the minute. Experiences like these can cause physical changes in your body and crank out stress hormones that have an impact on brain aging.

But don't send the kids to boarding school or fire the contractor. Not yet, anyway. Stress and anxiety can be reduced and even eliminated from our lives, and there are many approaches to help us achieve this. We can choose from a variety of stress-reduction methods, including yoga, jogging, meditation, prayer, or even anger-management classes.

What Causes Stress?

Both external and internal events can trigger a stress response. In our physical environments, we are constantly being bombarded by stimuli such as noise, bright lights, heat, or confined spaces—all of which can bring on stress, if the intensity and timing are right. People and various social situations can create stress, whether it's a rude waitress, critical boss, or a crowded rock concert or amusement park. Anything from a physical examination to a hospital visit, a public verbal presentation or just getting into a heated discussion, can activate our stress response. Deadlines, on and off the job, are common reasons for getting our adrenaline pumping, as are major life events, whether a negative one—death of a relative, getting fired—or even a positive one—getting promoted or having a new baby. And, let's not forget the daily hassles of commuting, misplacing keys, or a mechanical breakdown of one of the pieces of equipment so many of us have accumulated to help us eliminate stress.

Perhaps our most exasperating and harmful forms of stress come from within ourselves—our so-called internal sources of stress. Examples include the overloaded schedules we take on, the caffeine we drink, and the sleep we deprive ourselves of.

Justin G., the youngest partner in his high-powered litigation firm, attached his heart monitor and prepared to begin working out in his customized super gym, tucked into the loft of his town house. He had over $20,000 of the latest home gym equipment available, and his perfectly toned body "in less than an hour a day" was the proof. Starting with a level 6, random uphill run on his knee-saving treadmill, with the financial network flashing across the wide flat screen hung before him, he watched his heart rate monitor climb to the optimal level. While increasing his level to 7, the TV, treadmill, lights, and music went dead. He screamed in frustration over his "frigging fuse box" but it was worse. Power was out in the whole neighborhood.

Justin grabbed the phone but then slammed it back into the cradle, yelling, "Damn electric phone system!" He used his cell phone to call the power company, but the circuits were busy. He stomped and cursed and flipped the fuses again and again. His heart was racing and he was sweating, but he was feeling frustrated and angry instead of exhilarated by his usual post-workout endorphin high.

He finally got the power company on the phone and ranted and raved about his limited time, his exercise machines, etc. The woman on the phone said they were doing all they could and suggested that he try to relax and not get stressed out about it. Incensed, he yelled that his stress level was none of her business. She apologized and wished him well with it.

Glancing out the window, Justin noticed an electric company truck parking across the street a few houses away. He leaped down the stairs and out the front door, practically mowing down his neighbor Rob, who was stretching out before his morning jog.

Rob asked, "Hey, Justin, you OK?"

Justin, distracted, replied, "Yeah, man. I've got to get to the office, my power's out, and I haven't worked out . . ."

"So, want to go for a run?" Rob began running in place.

Justin, already across the street and about to chew out the electric company guys, looked back at Rob incredulously. "What? No. I need these jerks to turn my power back on."

Rob smiled as he jogged away. "Have a nice day, Justin."

Although much of the stress we experience on a daily basis is self-generated, most people think only of external stressors when they become upset. If only my boss, kids, or spouse would do things differently, then I wouldn't be so stressed out. If the stock market would just bounce back, I could relax. Coming to terms with our own personality and how we can begin to eliminate some stress for ourselves is an important step in minimizing our anxiety.

High-achieving baby boomers, sometimes described as Type A personalities, perfectionists, or workaholics, often set unrealistic expectations on themselves and over-analyze or worry about what other people think about them and their work. Others tend toward self-criticism and pessimism. These and other mental states or personality styles may cause people to be prone to anxiety, stress responses, and the accompanying release of stress hormones.

How Stress Affects Us and How We React to It

Stress is the body's response to a demand made upon it, requiring the body, mind, or both to adapt. This demand can take the form of a threat, challenge, or simply an unexpected change. Stress responses are usually immediate and automatic. Everybody responds to stress differently, and these responses are not always negative. A college student stressed out about exams may study harder and perform better, whereas another student may become overwhelmed by anxiety and freeze up at the exam.

Our bodies respond to stress by releasing hormones into the blood stream that are intended to put us into a protective mode. Adrenaline, a well-known stress hormone, tends to result in the "fight-or-flight response," providing strength and energy to either fight impending danger or escape it. This physiological response has been genetically programmed through evolution, perhaps dating back to our caveman ancestors. If another caveman came along to steal your food and you were a bigger guy, perhaps even had a big club, you might have fought him off. Otherwise, you'd probably have run for it.

When the stress hormone adrenaline pumps into the blood stream, heart rate quickens, blood pressure rises, and breathing gets faster. More blood and oxygen get to the heart, muscles, and brain. Muscles tense in preparation for action, mental alertness increases, sensory organs become more sensitive, and less blood goes to the skin, digestive tract, kidneys, and liver, since those organs won't be needed as much during a crisis. Sugar, fats, and cholesterol increase in the blood for additional energy, and platelets and blood-clotting factors increase to prevent bleeding in case of injury. All these physiological changes help us adapt to the acute situation our bodies believe to be at hand.

An adaptation to conditions requiring a rapid reaction, the human stress response evolved as a protection against acute threats and sometimes made the difference between survival and death. Unfortunately, this same physiological response can occur in people who are not exposed to physical threats but instead to constant or repeated mental triggers or stressors that have no rapid resolution. They persist and linger and smolder, leading to a chronic stress syndrome characterized by a variety of physical and mental symptoms, sometimes leading to health problems.

Common Symptoms of Chronic Stress

- **Physical:** headache, fatigue, insomnia, muscle aches and pains, rapid heart rate, chest pain, upset stomach, appetite loss, trembling, cold hands and feet, sweating.
- **Emotional:** depression, tension, anxiety, anger, frustration, worry, fear, irritability, impatience.
- **Mental:** poor concentration, memory loss, indecisiveness, confusion, poor sense of humor.
- **Behavioral:** fidgeting, pacing, nail-biting, foot-tapping, overeating, smoking, drinking, drug abuse.

The Effects of Stress on Memory Ability

Dr. Robert Sapolsky at Stanford University has studied how stress influences the brain and cognitive processes, showing that prolonged exposure to stress hormones has an adverse, shrinking effect on the hippocampus memory center in laboratory animals. The hippocampus is a seahorse-shaped brain structure involved in memory and learning, located in the area of the brain beneath the temples.

Dr. James McGaugh of the University of California at Irvine has shown that corticosterone, a hormone released by severe stress, anxiety, or even a physical blow to the body, can block the retrieval of information stored in long-term memory. His research group, using laboratory rats, found that a small electric shock elevated corticosterone, crippling the animals' ability to find their way back to a designated target. Their memory was impaired most while the hormone levels were at their highest, up to an hour after the initial shock. Although the memory loss in this experiment was temporary, it raises questions about the long-term effects of repeated stress on the brain.

Dr. John Newcomer of Washington University School of Medicine in St. Louis observed similar stress effects on memory in humans. His group showed that several days of exposure to high levels of the stress hormone cortisol can impair memory. The scientists observed memory impairment only in people treated with high doses—comparable to what a person would experience after a major illness or surgery. A week later, however, their memory performance returned to normal. Although these results suggest that only people who experience severe medical, physical, or psychological trauma will experience stress-related memory impairment, many researchers are convinced that long-term exposure to lower stress levels is also likely to accelerate brain aging.

Getting Mad, Sad, or Even

When faced with a frustrating, seemingly unsolvable problem, our emotional reactions can vary considerably. Anger, fear, sadness, and denial—sometimes expressed as humor—are common responses, and each of these feelings has positive and negative consequences. How our emotions motivate us to act has a significant impact on keeping our brains young.

Some people tend to become overwhelmed by angry feelings or are unable to express them. As a result, they may not fight for their beliefs, try to get even, or even get satisfaction, but instead give up—a response sometimes leading to sadness or even depression.

Sonia J., a 70-year-old widow, was becoming increasingly frustrated and anxious about her constant forgetfulness. It wasn't so bad at first, but now she was reminding herself of her exasperating older brother, Marty, who was constantly forgetting things and driving everybody crazy. Sonia's son was concerned that perhaps she was depressed and that maybe she should see a therapist. Sonia dismissed the idea completely—she didn't buy into all that "talk it to death" stuff.

At Sunday dinner, Sonia snapped so angrily at Marty for again repeating something he'd just told her that she started shaking and had to go lie down. She realized later that part of why she got so mad at Marty was because she was afraid about her own forgetfulness. The next day she called her doctor for help.

Following an evaluation, Sonia's physician began her on an anti-Alzheimer's drug, and after several weeks her memory showed improvement. Sonia was happy and relaxed for the first time in months. Her son was pleased and relieved.

Sonia tried to get Marty to take the same medicine, or at least talk to his doctor about it, but he vehemently denied any memory problems and refused Sonia's urgings to see any doctors or take any medicines. Unfortunately, Marty's memory declined rapidly during the next twelve months when he was eventually diagnosed with Alzheimer's disease.

Anger can sometimes motivate us to act in a positive, constructive manner. However, anger, expressed or not, can also lead to high levels of anxiety, stress hormones, depression, and even memory loss.

It doesn't necessarily serve us to express every feeling or emotion, especially if we alienate relatives or colleagues, and because we may feel differently once the angry feelings have diminished. Sometimes it's better not to zoom down to the post office at midnight in your pajamas with that irate letter to your ungrateful, demanding boss until you have read it over again in the morning. The stress of unemployment may prove more detrimental to your brain fitness, not to mention your pocketbook, than the momentary thrill of yelling "I quit" in the boss's face. Anger management therapy techniques involve learning to understand our feelings and finding new approaches for expressing them.

A recent study supports the idea that outright expressions of anger may not always be the healthiest solution to stress. Dr. James Blumenthal and his associates at Duke University studied the effects of anger, as well as those of physical exercise, on heart disease. They found that the group of cardiac patients who exercised *and* learned anger management techniques ended up with the lowest risk for ischemic chest pain, resulting from insufficient blood flow and oxygen to the heart. The group of patients who only exercised without anger management instruction lowered their risk for chest pain slightly, but only half as much as the other group. The physiological process leading to this kind of circulatory problem in the heart has the potential for producing similar circulatory problems in the brain.

Dr. Frank B. was a busy internist on the Upper West Side of Manhattan. Although 61 years old, he was in good shape and looked younger. He prided himself on keeping up with all his journals and latest medical advances while continuing to practice like an old-style family doctor who still made an occasional house call.

During the last four years, he had noticed his memory gradually declining. He used to know all of his regular patients' main health problems and the names of their close relatives.

Lately he had to refer to their charts—sometimes even for basic facts and details. On several recent occasions, if Frank bumped into a patient at a local restaurant or bookstore, he would recognize their face but couldn't recall their name until hours later. This began to secretly terrify him.

Frank was five years into his second marriage with Patricia, now in her mid-forties. He was afraid to let her know the extent of his memory concerns—after all, he didn't want to worry her. However, she had already noticed on her own, as did some of his staff, closest friends, and patients. When Patricia brought it up, Frank insisted he was fine—and he should know, he's a doctor, a healer. He doesn't need to be healed by anybody else, thank you.

His practice began to suffer. Frank was seeing up to five patients an hour and was still expected to dictate notes between appointments. He started having stomach pain and headaches and couldn't keep up.

Patricia finally dragged him to a neuropsychiatrist specializing in memory problems. When his physical exam and lab tests turned out normal, Frank grinned at Patricia, vindicated at last. However, the doctor went on to explain that Frank did have subtle memory losses and a strikingly high level of anxiety and stress, which was most likely aggravating his head and stomach pain, as well as his concentration and memory.

Over the next few months, Frank made some changes in his lifestyle. He brought a young partner into his practice, which greatly reduced his patient load and stress at work. He began playing tennis again with friends and accompanied Patricia to her yoga class twice a week.

People at the office began remarking on how alert and focused he seemed, and Frank himself noticed that his memory for names and details had improved. Also, he had fewer aches and pains and he felt a general sense of well-being. He and Patricia felt closer than ever. Frank wasn't sure which stress-reducing change had done the trick, and he didn't care—he just stuck with the program.

Stressing Down, Not Stressing Out

We are often powerless to affect or reduce the external stressors in our lives. However, because much of our daily stress is internal or self-generated, we have the ability to do something about it, if we choose.

People who are able to maintain positive outlooks on life may actually live longer. The most recent finding from *The Nun Study* indicated that the nuns who expressed feelings of joy, happiness, love, and hope in their early diaries lived as much as ten years longer than those who were less positive.

If time allowed, I would love to write a book called *The Relaxation Bible: An Innovative Strategy for Stress Reduction*, but I don't think I could tackle the added stress right now. In the meantime, the strategies contained in the following pages should prove useful in reducing your stress and thereby improving your memory ability.

SET REALISTIC EXPECTATIONS

Many of us set unrealistic expectations upon ourselves as well as others, and although this is a frequent source of self-inflicted stress, it is one of the easiest to change. You can't become a concert violinist or marathon runner overnight—it takes years of practice and hard work. If you buy a violin with the goal of playing Stravinsky in two weeks, you will put yourself under tremendous stress and most likely fail. When expectations become more reasonable, we gain a sense of control in our lives and are able to plan and prepare ourselves both physically and psychologically.

LET'S GET PHYSICAL

Recent studies indicate that physical exercise improves memory function (Chapter 8). It may also reduce stress through its release of endorphins, the body's natural antidepressant hormone. In a sense, exercise works off much of our "stress energy." With the fight-or-flight reaction that adrenaline and other stress hormones bring about, many of us are inclined to either act somewhat impulsively or keep these hormones and impulses bottled up inside. In today's modern world, fight or flight—though it worked for cavemen under attack—may not be an option. Our bodies thus remain in a state of heightened energy with no release.

Exercise helps us dissipate such excess energy. Channeling the energy into a brisk walk will be more effective in reducing stress than gulping down a couple of beers. Any aerobic activity can have the same effect, whether it's jogging, walking, bicycling, swimming, racquet sports, or aerobics classes. Be sure to choose activities you like and vary your exercises to maintain your interest.

Approximately 15 million Americans include some form of yoga in their fitness program. Yoga not only offers a way to build strength, balance, and stamina, but it can also reduce stress. Dr. Dean Ornish and his co-workers found that 80 percent of the cardiac patients in their experimental group practicing yoga along with other lifestyle interventions were able to avoid coronary bypass surgery.

PREPARE AHEAD

Much of our stress and anxiety depends on the situation we are in. Many people fear public speaking, whether it's a sales pitch, a marriage proposal, or a State of the Union address. I am struck by how many people fear that they will get up there and forget everything they are supposed to say and do.

Any new or unfamiliar situation can create stress and anxiety, particularly if we face it unprepared. Therefore, an effective approach is to prepare in advance. When I studied piano as a teenager, my father always advised me to practice "110 percent" so I had 10 percent "cushion" room left for anxiety, fear, and forgetfulness. I still flubbed a few notes at recitals, but I understood his point and it served me well.

When preparing for an oral report, speech, or exam, you might visit the location in advance, if possible, and familiarize yourself with it. Taking slow, deep breaths, as well as closing your eyes and envisioning a calm place just before a performance, can help you to relax when feeling anticipatory anxiety or stress. Another useful technique when speaking in public is to focus on one person in the crowd. Some people find it helpful to take their feelings of anxiety and rename and transform them into feelings of excitement and energy. Others like that old trick of imagining their audience in their underwear. I guess it all depends on your audience.

RECESS ISN'T JUST FOR KIDS

Most of us lead fast-paced lives and pay little attention to warning signs that it is time to rest. For some, a certain mild, optimum degree of stress can lead to a healthy tension that helps them function at their best. However, excessive stress, or *distress*, can cause fatigue and eventually exhaustion—glaring red flags telling us to slow down and rest. Pacing ourselves while we work and play involves monitoring our levels of stress *and* energy, and taking breaks when we need them, much like small children need their naps to behave nicely.

A helpful approach to avoiding the workaholic syndrome is to take periodic time-outs. For many of us, breaks are built into our daily schedules, as we tend to divide our days into four 2-hour segments: mid-morning break, lunch, mid-afternoon break, and dinner. We can use these times for power naps, meditation, yoga breaks, walks, refreshments, and other activities that recharge our emotional and physical batteries, increase productivity, and reduce stress levels.

RELAX, THIS WON'T HURT

Whether you practice yoga every day, meditate, or sing in the shower, any conscious effort you make to relax, both mentally and physically, will reduce stress. Dr. Herbert Benson of Harvard University has described this process as the *relaxation response*. Just as our bodies evolved and developed an automatic stress response, we can teach ourselves, through conscious effort and repetition, to switch on a relaxation response—a state of deep mental and physical relaxation. Physiological activities slow down—heart and breathing rates decrease, blood pressure lowers, and muscles relax.

Such simple activities as resting at the beach, lying on a favorite hammock, or cuddling up with a good book can bring about this state. Also, just *imagining* resting at the beach or in your favorite hammock while taking a couple of minutes of quiet, deep-breathing time at work can have a similar stress-reducing relaxation effect. Deep relaxation can be accomplished through a variety of techniques, including yoga, tai chi, biofeedback, meditation, and self-hypnosis, all of which can be learned through courses, books, and tapes. Just a few minutes each day doing some simple relaxation techniques can be effective in helping us to remain calm and perform at our optimal memory capacity (see Relaxation Exercises box).

Relaxation Exercises

- ***2-Minute Break.*** Lie down or sit in a comfortable position. Begin by breathing slowly through your nose, regularly and deeply. Focus on your rib muscles, expanding them as much as possible, then slowly pushing out as much air as possible. Be sure to use your diaphragm, and keep your breathing deep, slow, and calm. Feel your abdomen rise as you breathe.
- ***5-Minute Break.*** Close your eyes and imagine yourself in a calm, soothing setting—at the beach, in a field, in a sauna, or anywhere you find relaxing. Breathe deeply and do not allow the thoughts that may enter your mind to remain there. Keep focused on your breathing and relaxed setting.
- ***10-Minute Break.*** Sit in a comfortable chair or lie down. Close your eyes and take a deep breath; let it out slowly. Focus your attention on your head and scalp, and then imagine releasing all the tension there. Bring your focus down to your facial muscles and release that tension. Let that relaxed feeling extend through your cheeks and jaw. Slowly continue this process, focusing down your neck and shoulders, releasing the tension and continuing to move systematically down your body through your arms, hands, abdomen, back, hips, legs, and toes. Continue to breathe deeply and slowly throughout.

CUT BACK ON CAFFEINE

Many of us have the caffeine habit, and we tend to get the bulk of our daily caffeine from drinking coffee. We may start with a wake-me-up cup, possibly followed by a mid-morning espresso and perhaps an iced blended mocha at our afternoon coffee break. If we count added caffeine from soft drinks and chocolate, we can be well on the road to a caffeine-induced stress response.

When caffeine levels go beyond what the body will tolerate—and this toleration level diminishes with increasing age—symptoms of stress and anxiety emerge. You may say that caffeine helps you focus and maintain attention, and in small amounts it can. However, at higher levels, caffeine can cause irritability and distraction.

I recommend cutting back on caffeine and doing it gradually to avoid headaches and other side effects of withdrawal. Many experts recommend decreasing by the equivalent of a half-cup of coffee each day or every other day (see Chapter 7 for equivalencies). Most people will begin feeling more relaxed and notice other benefits as well. Many find that they sleep better and paradoxically have *more* energy.

GET ENOUGH SLEEP

An estimated 100 million Americans do not get a good night's sleep on a regular basis. Throughout the world, an even larger number of people live in a chronic state of sleep deprivation. Sleep-deprived people rarely awaken refreshed each morning, and they lack energy during the day. The average person needs about seven to eight hours of sleep each night, though our need for sleep decreases with age. Getting enough sleep is essential for normal brain development. Studies of laboratory animals indicate that adequate sleep enhances the connections between brain cells.

Insomnia and fatigue are major sources of stress that can impair concentration and memory. When sleep patterns improve, so do mood and memory. People who suffer from chronic sleep deprivation often feel better if they try getting to bed 30 to 60 minutes earlier. You know you've beaten the cycle if you start waking refreshed, notice more energy during the day, and find yourself waking naturally before the alarm goes off in the morning.

Those weekend days when you can sleep in may help you recover from chronic sleep deprivation, but if you sleep too long your body rhythms may get thrown off the next day. Daytime naps can help if you keep them short—you could feel groggy waking from naps that last over 30 minutes. Instead, 20-minute "power naps" can be rejuvenating. Avoid early evening naps since they make it more difficult to fall asleep at bedtime. If you suffer from chronic insomnia, avoid daytime naps altogether, and try a systematic sleep inducement program instead (see box). Sometimes chronic insomnia is a symptom of depression or some other medical condition, so consult your physician if a sleep inducement program is ineffective.

Beat Insomnia at Its Own Game: A Systematic Approach to Sleep Inducement

1. What to avoid:

   - **Daytime naps**
   - **Evening liquids**
   - **Exercise or excitement an hour before bedtime**

2. Begin your sleep inducement program on a weekend, preferably a Friday night.

3. Get into bed the same time each evening. Once in bed, do not watch TV or eat or even read a book—just turn out the light, get yourself in a comfortable position and relax (see earlier Relaxation Exercises box).

4. If you are not asleep after 20 minutes, get out of bed and do something else: watch TV, listen to music, or read a book.

5. Once you begin feeling tired, go back to steps 3 and 4: go to bed, shut the light, relax. If you're not asleep after 20 minutes, get out of bed and do something else.

6. Do not worry if you spend a good part of the night out of bed. A key step to the program is avoiding naps the next day. If you can manage to stay awake the next day, you will likely conquer your chronic insomnia in just a few days. The next night, your fatigue will kick in at bedtime (make sure it is a consistent time). Go back to steps 3 and 4 and continue to avoid daytime naps.

BALANCE WORK AND LEISURE

Although new technologies and devices help us save time and energy, Americans on average work about three hours longer every week than they did twenty years ago, adding up to an extra month of work each year. With many boomer couples pursuing two careers, their family and leisure time becomes even scarcer. The word *leisure* comes from the Latin word *licere*, meaning permission—we need to give ourselves permission to take our leisure time and enjoy life. People who never allow themselves the leisure time they need experience greater levels of stress.

To tally the balance of work and leisure in your life, take out your notebook and add up the number of hours you spend in each area throughout the week, not including sleep. If you spend more than 60 percent of your week at work or doing work-related activities, you probably need to think about shifting the balance more toward leisure. We all need time for exercise, relaxation, socializing, entertainment, and hobbies, and this leisure time will reduce stress. Some of us resist taking "personal time" because it makes us feel guilty or selfish, or too much leisure time makes some people bored and restless—even stressed! We need to find our own balance of leisure time versus work time, one that allows us to limit our stress level, and maintain our optimal memory ability.

LAUGH YOUR HEAD OFF

Humor, too, can reduce stress—it puts uncomfortable feelings into perspective, giving us greater distance from them and releasing emotional discomfort and pain through the pleasure of laughter. Norman Cousins advocated the use of humor not just to reduce stress but also to cure physical ailments, through the physiological effects of laughter. Although laughter does relieve tension, it has not been proven to cure physical illnesses, yet. However, I have never met an ill person who has complained to me about laughing too much.

TALK ABOUT FEELINGS

Whether you call it venting or getting emotional support or letting it all hang out, there is no question that talking about feelings is one of the most effective ways to reduce stress. Systematic studies of talking psychotherapies often find that the characteristics of the listener—whether they are empathic, responsive, and the like—have more importance to the therapeutic benefit than what type of therapy they practice. Talking about feelings can be effective with a spouse, sibling, parent, bartender, mah-jongg partner, psychiatrist, priest, or any of a number of people who make you feel comfortable. The experience sometimes leads to tears, a sense of relief, and, when the person listening does not judge or criticize, a sense of understanding and acceptance. It puts the troubling feelings into perspective, making us feel strength and distance from whatever the source of stress.

Review of Steps to Minimize Stress and Anxiety

- **Set realistic expectations.**
- **Exercise regularly.**
- **Prepare ahead.**
- **Take breaks throughout the day.**
- **Learn how to relax and do it at regular intervals.**
- **Cut back on caffeine.**
- **Get enough sleep.**
- **Balance work and leisure.**
- **Let yourself laugh.**
- **Talk about feelings.**

When Stress Becomes Chronic: Depression and Anxiety

Sometimes, despite our best efforts to reduce stress, we are unable to eliminate the sources of tension or emotional pain in our lives. Sometimes we may go from directing our anger at the outside world and instead turn it within, becoming mad at ourselves. This mental process tends to change from anger to sadness and, eventually, to depression.

An estimated 15 percent of the population develops an episode of depression requiring medical intervention at some point in life, and stress is not the only cause. Some people are born with a biological predisposition to get depressed, tending toward a brain chemical imbalance that favors a depressed mental state. In some situations the depression appears to come from nowhere, even in the best of circumstances. Perhaps more frequently, the depression stems from a combination of stressful life events and internal biological factors.

Depression and Memory Loss

I have seen many patients become worried and depressed about their objective or subjective declining memory abilities. This syndrome can develop into a vicious cycle in which the worry over memory loss deepens the depression, which, in turn, increases the forgetfulness or memory loss. The situation can become exacerbated when it triggers the concern and anxiety of family members. Sometimes there are people around us to guide us toward help, other times we must seek help for ourselves.

Holly M. awoke from another mid-afternoon nap. She took a nap almost every day now, ever since their youngest daughter left for college. It was half past five and Carl, her husband of thirty-six years, would be home in an hour. She had time to shower and dress and fix him a T-bone, charred outside but pink inside—just the way he liked it. But wait . . . didn't she make steak yesterday? Oh, never mind. Carl always ended up working late anyway, so it didn't matter.

At 8:00, showered, dressed, and made up perfectly, Holly sat at the dining room table as the steaks grew cold and she longed to escape back to sleep. At 10:00 Carl finally came in and was shocked to see her sitting there like a zombie. He yelled, "What the hell is going on?"

"You could have called if you were going to be late for dinner again," she said with a sarcastic clip.

He laughed quietly and shook his head. "This memory lapse thing of yours is getting out of control, Holly. I specifically told you, *this* morning, that I had to have dinner in the city with clients. Maybe you should see somebody. A shrink or something."

Holly's eyes teared. She asked him what was happening

to her. Why was she forgetting so much? Why couldn't she concentrate on anything? Why did she feel so sad all the time? She stared at him, pleading for help.

Carl was cold. "I have some calls to make. Why don't you just go to bed and get some rest." After closing himself in the study, Carl made a call and a woman answered. He said he had a wonderful evening and he'd like to see her again—tomorrow? She inquired about his wife, but Carl just laughed and said she didn't remember a thing from minute to minute and not to worry.

The next day Holly decided to take Carl's advice and got a referral for a psychiatrist. At the first appointment, she learned that she was depressed, and this could account for some of her memory problems. The doctor prescribed an antidepressant medicine and suggested they meet a few more times to talk about her feelings. Holly left his office feeling relief and optimism.

Over the next couple of weeks, Holly's mood improved along with her memory. Unfortunately, Carl was so seldom at home, he didn't notice. Friday morning Carl came down to the kitchen with a packed overnight case and headed for the door. Holly, already dressed and about to leave for the gym, blocked his path. Carl raised his eyebrows. "Good for you, Holly. Get some exercise. It might help you feel better. See you Sunday night!" Holly was taken aback. "What? Where are you going?"

Carl snapped, "You don't remember this either? I told you on Tuesday! I've got a statewide sales meeting at the regional office all weekend! For God's sake, Holly, get a grip! I'll call you."

At this point, Holly had a pretty good grip. Something was amiss, and it wasn't her memory. Now that her depression had lifted and her memory had improved, Holly took the initiative to figure out what was going on in her marriage. She discovered her husband's affair and realized he had been using her own symptoms against her, tricking her into believing his lies.

Holly ended up with the house, the money, and the cars, but she let Carl keep his videotape collection, including Alfred Hitchcock's classic *Gaslight*.

Antidepressant drugs can have a major impact on depression (Chapter 9), but talking therapies can be powerful interventions as well. Dr. Charles Reynolds and his associates at the University of Pittsburgh studied a group of depressed patients who had been successfully treated with antidepressant medications. The researchers followed these patients for an additional twelve-month period while one-third continued taking antidepressants, one-third took placebo medicine, and the last third received psychotherapy. Dr. Reynolds found that only 20 percent of the subjects who stayed on antidepressant drugs had a relapse of their depression, while 80 percent of those who took a placebo became depressed again. By contrast, only 50 percent of the patients who received psychotherapy experienced another depression during the follow-up period, a clear and striking benefit over that of the placebo. What is remarkable about the study is that the psychotherapy involved only one hour a month, a far cry from in-depth psychoanalysis. For some forms of depression, then, brief monthly meetings with a therapist may be enough to relieve symptoms.

While many people have a predisposition to get depressed when under prolonged stress, many others tend toward anxious states. And still others experience mixed states of anxiety and depression. Anxiety disorders can be disabling and come in many forms. Panic disorder is a condition involving intense, sudden attacks of anxiety and can often evolve into agoraphobia, wherein a person may avoid the situations associated with the attacks. These avoidance patterns can progress to the point where a person becomes housebound. Obsessive-compulsive individuals experience unwanted obsessive thinking and impulses, which can lead to compulsive behavior like washing hands or checking door locks over and over. These disorders can become so severe that the afflicted are unable to function in their lives. Some people experience a pervasive, continual generalized anxiety or even more focused fears and phobias, which can pervade all aspects of their lives.

These conditions result from both external stress and internal biological factors. Regardless of the cause or form of anxiety, these conditions often respond to drug treatment as well as specific psychotherapies. And, many such psychiatric disorders will affect learning and memory abilities.

If you find that your anxiety levels are so high that your work or personal life is affected, perhaps it is time to seek professional help (Appendix 5). The stress reduction techniques mentioned earlier may help to some extent, but severe anxiety and panic states can be just as debilitating and dangerous as extreme depressions.

Chapter Five

Get Fit with Mental Aerobics

Man's mind, once stretched by a new idea, never regains its original dimensions.
—Oliver Wendell Holmes, Jr.

Jill S. had never been very good at remembering names. In her early forties, this lifelong difficulty took a turn for the worse. Because of her busy career in marketing, Jill had put off having children into her late thirties and was now in the heat of carpool years. With her kids' soccer practice and ballet lessons, and her husband's frequent business trips, Jill was at her wit's end trying to juggle everybody's schedules, let alone keep the names of new business contacts straight. When she came to my office, Jill was exhausted and frustrated. She was becoming forgetful, and for the first time in her life, she was afraid things were going to fall through the cracks and she was going to "blow it."

Jill's story is similar to that of millions of baby boomers who are creeping into middle age. She wanted to take action to improve her memory now and organize her life more efficiently. She is one of many proactive individuals eager to get fit and benefit from a mental aerobics program.

Mental aerobics is any mental activity that exercises your brain. Just as sit-ups tighten your "abs," mental aerobics are jumping jacks for your mind. Just picture a mini Jack La Lanne in your brain. You don't remember him? Drop and give me 20!

In the memory-training section, we learned mental tools for improving learning and recall with the goal of practical daily use. Initially these techniques serve as mental aerobics, in a sense increasing the stamina and strength of our brain cells. Once mastered, they become routine and helpful in our lives, but we still stand to benefit from a daily regimen of mental aerobics, which continue to challenge us mentally and keep our neurons firing in top form. Just as joggers gradually lengthen the distances they run over time to increase their aerobic workout, we need to increase the complexity of our mental aerobics program, whether it's doing crossword puzzles, solving brain-teasers, playing charades, or watching *Jeopardy!*

The Mozart Effect

Educators have observed that young children, from toddlers to preteens, who are exposed to Mozart compositions and other classical music, appear to perform better academically than those who are not. In studies of college students, Dr. Francis Raucher, Dr. Gordon Shaw, and other neuroscientists at the University of Wisconsin showed that listening to a Mozart piano sonata improved the students' cognitive abilities. Interestingly, it was not verbal or language skills that improved but rather spatial cognitive skills, such as paper-folding tasks and following patterns. The researchers speculated that listening to music helps to temporarily organize thinking and that mental processes involved in listening to music activate a neural network that is shared with spatial-reasoning processes. Other investigations have found that some college students perform better on cognitive tests when they take the tests with background classical music instead of silence.

Although the concept of the Mozart Effect has met controversy since not all studies show it, we do know that different kinds of music have different mental effects. Some music will calm us, lowering heart rate and blood pressure, while other musical styles are likely to agitate us. There is evidence that listening to music can enhance immune function and diminish pain. Several experts believe that the logic, symmetry, and aesthetic organization of classical pieces by Mozart, Beethoven, and others truly provide a mental advantage to people of all ages.

Dr. Gottfried Schlang and Dr. Gaser Christian of Beth Israel Deaconess Medical Center in Boston recently used MRI scans to study whether intense environmental demands such as musical training at an early age influenced actual brain growth and development. They found that, compared with non-musicians in their study, the fifteen professional musicians had significantly greater volumes of gray matter—the outer part of the brain that contains the nerve cell bodies. The gray-matter areas showing the largest relative size were those involved in sensation, motor function, and one of the areas involved in memory function that is affected early by Alzheimer's disease. Though the evidence is circumstantial, it is consistent with the possibility that musical training in early life could offer protection against Alzheimer's disease later in life.

We know music can elevate a person's mood, and a better mood certainly can sharpen mental ability—a depressed person is often distracted and unable to focus on mental tasks. Even without definitive proof, the potential for a benefit and the minimal risk involved convinces me that listening to classical music may be a worthwhile habit for us all.

Use It or Lose It

Charles W., a 47-year-old newspaper journalist and father of two, was a volunteer for one of our UCLA memory studies. He was having minor trouble remembering facts and background details of his feature stories, eventually requiring him to make at least twice as many notes as usual to get his stories down. Interestingly, during college he had a passion for crossword puzzles, but had given it up for lack of time. While consulting with me on his memory improvement program, I suggested he take up crosswords again for the potential mental aerobic benefits. He became a voracious puzzle solver—after six months he was completing the Sunday *New York Times* puzzle in ink using a stopwatch. His crossword accomplishments gave him confidence, and his memory on the job improved.

A PhD in engineering is no guarantee against developing Alzheimer's disease—the disease strikes people from all walks of life, including ex-presidents, Nobel laureates, and nuclear physicists. Recent studies, however, indicate a definite link between mental activity and staving off symptoms of Alzheimer's disease. And both laboratory and clinical studies point to the memory benefits of mental activity.

Researchers at Case Western Reserve University found that the risk for developing Alzheimer's disease was three times lower in people who had been intellectually active during their forties and fifties compared with those who had not. Their diverse mental activities included reading, working jigsaw puzzles, woodworking, painting, knitting, and playing board games. Couch potatoes didn't reap the benefit: passive pursuits like going to the movies did not contribute to the lowered Alzheimer's risk. Of course, watching an enjoyable movie may reduce stress, which has its benefits for memory (Chapter 4), but beware of over-salted popcorn (Chapter 7).

Rush University researchers found higher mental stimulation in one's twenties predicted better cognitive function late in life. People who spent time reading and had mentally stimulating jobs or educational experiences maintained their memories better and longer as they aged. As mentioned previously, other research groups have shown that college graduates have a lower risk of eventually developing Alzheimer's disease than people with less educational achievement.

At UCLA we have recently added another dimension to this line of research by exploring whether educational achievement protects against brain aging. We used PET scanning to determine whether prior educational achievement, such as a four-year college education, is associated with higher brain activity levels in people with normal memory abilities (Chapter 1). The results confirmed our prediction based on the earlier population studies: college graduates had higher activity in a critical part of the brain, the posterior cingulate, which is involved in memory performance.

These observations point to the "use it or lose it" theory: people who use their brain cells will keep them fit and protect the cells from "wear and tear." These neuron users may be less likely to "lose it" as they age. Of course, the PET scan may be showing us a healthier brain that was healthier at birth, and our genetic predisposition for a healthier brain may have also gotten us on the college trajectory to begin with.

Neuroscientist Fred Gage and colleagues at the Salk Institute added weight to the "use it or lose it" theory in studies of newborn rats in enriched environments with treadmills, toys, and a variety of foods. The rodents in the enriched environments had significantly more neurons in their hippocampal memory centers compared with rats living in ordinary laboratory cages. Dr. William Greenough's research team at the University of Illinois found that rats in more stimulating environments grow new brain cells, more synapses, or communicating connections between the cells, and new blood vessels for transporting oxygenated blood to feed their more active brains. And, when running through their mazes and completing other memory tests, the stimulated rats appeared more intelligent.

Additional research supports the idea that continual, lifelong mental stimulation is healthy for human brains as well. Mentally and physically active people over age 65 have been found to have higher IQ test scores and higher blood flow in the brain compared with those who remain inactive over a four-year period. Not surprisingly, people with advanced education and professional accomplishments tend to have greater density of neuronal connections in brain areas involved in complex reasoning.

Other recent research from Case Western Reserve University has provided an additional interesting observation: people with mentally demanding jobs—managers, professionals, and so forth—experience less memory decline as they age when compared with their counterparts who have less demanding jobs. It must be noted, however, that although this observation agrees with the "use it or lose it" theory, people destined to develop Alzheimer's disease may be predisposed to choose less demanding occupations in the first place.

These discoveries point to the conclusion that mental stimulation, or exerting our brains in various ways intellectually, may tone up our memory performance, protect us from future decline in brain function, and may even lead to new brain cell growth in the future!

Cross-Train Your Brain

Fitness trainers often advise their clients to cross-train, or vary their workout and avoid repeating the same exercise routine day after day. Cross training challenges athletes, minimizes boredom, and maximizes results. Also, varying one's workout by focusing on a particular muscle group one day and a different muscle group or activity the next allows an athlete to rest muscle groups between workouts, which builds stamina.

Neuroscientists believe the same principle holds true for brain training. Dr. Arnold Scheibel at UCLA has described the way our brains thrive on novelty. Unfamiliar stimulation and new mental challenges actually stimulate growth in a section of the brain known as the reticular formation. This brain region may have developed its novelty-seeking specialization as a survival mechanism or adaptation to the need for our ancestors to spot predators.

Brain Training, Not Brain Straining

An old tennis partner once told me that he suffered from an "exercise disorder." He was a typical Type A personality who embraced every new exercise program he could find and bought any workout gadget late-night TV had to sell. His disorder was that he would overdo each program and quickly injure himself, which would force him to take an extended break from all exercise in order to recover. Afterward, he would promptly begin a whole new exercise regimen designed to heal the injury, but would eventually injure a new joint and repeat the cycle.

Just as in physical fitness training, in memory training we want to avoid too much of a good thing. At UCLA, we have found that people with the APOE-4 genetic risk for Alzheimer's disease have to work harder to memorize and recall the same information as people without the genetic risk. When research volunteers play a computer game for the first time, their PET scans reveal high brain-activity levels. After becoming proficient at the game, however, the brain scans show minimal activity during play. They need to use less of their brain capacity to accomplish the same task, much like an athlete becomes more proficient in lifting weights or running a marathon after training. This research points to the possibility that if we allow our brains to train gradually, just as a weight trainer increases the weight they lift gradually, it may be possible to accomplish the same performance level with less effort and frustration.

The scientific evidence points to mental stimulation and brain training as a way to maintain healthy brains throughout life. Suggestive evidence indicates that anything we do to exercise our brains in a *new* way may help to develop nerve pathways that can help to forestall the effects of Alzheimer's disease. Most of these approaches are inexpensive, not harmful, and certainly worth a try.

It is critical to begin mental aerobic exercises at a level that stimulates but does not over-exert. If a task is too difficult, a person may get frustrated and give up. If it is too easy, one may lose interest and get distracted. In our research using cognitive stress tests, we found that patients with even mild Alzheimer's disease were unable to perform the more challenging memory exercises—they became frustrated and lost track of the task. Rather than seeing brain activity in their memory centers, as we did with volunteers who had only mild memory complaints, we saw either no activity or else activity in brain emotion centers, probably reflecting their frustration from trying to complete an overly challenging mental exercise.

Brain Workouts Through Creative Thinking: Puzzles and Brain-Teasers

The information in our brains is passed through billions of dendrites, or extensions of brain cells, similar to branches of a tree, which grow smaller as they extend outward. Without use, our dendrites can shrink or atrophy; but when we exercise them in new and creative ways, their connections remain active as they pass new information along. And, remarkably, new dendrites can be created even after old ones die.

Evidence shows we can "work out" our dendrites and extend their branches in many ways. Even routine daily activities like lacing a shoe or rinsing dishes can be a trip to the gym for those little guys. Try tying your shoelaces backward or brushing your teeth with your left hand (if you're right-handed)—both could stimulate a neuron or two. Basically, any conscious effort to tease your brain can potentially create new brain cell connections.

The fun of solving puzzles and brain-teasers often comes from pushing ourselves to make a mental leap from existing assumptions to find a new solution to a seemingly unsolvable problem. To do so, we need to break loose and explore the problem, puzzle, or brain-teaser in a new way.

When we view certain visual images, we often fix on seeing them in one way, as in the vase below:

If you look again and think of the vase as background instead of foreground, you may see the profiles of two people.

In the figure below, you probably see the black arrows. Try to see the figure from a different perspective and push those black arrows into the background. Can you now see the white arrows emerge facing the opposite direction?

Sometimes our mental assumptions actually distort reality. Look at the figure below. Does the upper line appear longer than the lower one?

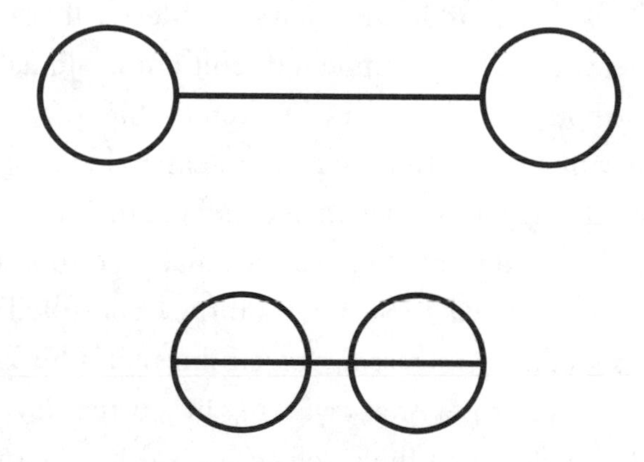

Take a ruler and measure the two lines and you'll see that they are of identical length.

The above exercises are basic examples of visual brain-teasers, the type that can and should become a part of your daily mental aerobics workout. You may want to go up to the attic and find your old Rubik's Cube. What is a Rubik's Cube, you ask? Drop and give me 50!

The goal of aerobically working out our brains is to get ourselves to think creatively in order to stimulate, strengthen, and enhance our brain cells, to maintain healthy dendrites and extend their branches.

Mental Aerobics: Getting Started

No fancy workout clothes, no expensive gym bag needed here. Your old comfy slippers and your favorite recliner will do fine. Your regimen of stretching, toning, and strengthening your brain can include music, puzzles, and computer games. Such activities are most effective when they not only are fun but they "shake up" your usual mental assumptions and force you to think of novel solutions.

Take a moment to review the results of your memory assessments from Chapter 2. These will point you to your optimal level to begin your workout. Also, take note of how you feel when you perform brain-exercise activities. If you find yourself getting frustrated quickly, go back and start at a less advanced level. If you find the activities too easy, move on to more difficult ones. Mental stimulation exercises should be challenging *and* enjoyable to achieve their best effect. Be sure to pace yourself and set reasonable expectations.

Many experts support the potential benefit of mental stimulation to our brains. But what form of mental stimulation is most effective? Recently, an experienced 52-year-old attorney consulted with me because of his gradually increasing forgetfulness and his family history of Alzheimer's disease. After reviewing his current level of mental activity, it was clear that the caseload he had been carrying had become stressful. The challenge for his brain fitness program was to *bring down* the level of mental stimulation in his life rather than add mental aerobics. In fact, we focused our discussion on ways for him to reduce stress in his life (Chapter 4).

My approach to mental aerobics is for each of us to identify a way to stimulate our brains without stressing them. The following includes a variety of mental aerobics exercises presented at different levels of difficulty that you can try out for yourself. As you familiarize yourself with them, you will be able to determine which level and type of exercise gives you a sense of mental stimulation without frustration. Once you know the kind of mental aerobics that works best for you, you may want to expand your repertoire by seeking additional resources on the Internet or at the library.

The following exercises are divided according to beginning, intermediate, and advanced levels, as well as which part of the brain each exercise trains. For most right-handed people, visual and spatial tasks work the brain's right hemisphere, while verbal or analytic tasks work the left hemisphere. For left-handers, the left side of the brain generally operates visual tasks, while the right side handles verbal skills.

LEFT BRAIN FUNCTIONS

- Logical analysis (reasoning, drawing conclusions)
- Information sequencing (making lists, organizing thoughts)
- Language and speech
- Reading and writing
- Counting and mathematics
- Symbol recognition

RIGHT BRAIN FUNCTIONS

- Spatial relationships (reading maps, doing jigsaw puzzles)
- Artistic and musical abilities
- Face recognition
- Depth perception
- Dreaming
- Emotional perception
- Sense of humor

Ideally, you want to work both hemispheres, and you may want to alternate your mental aerobic stimulation program from left hemisphere to right hemisphere.

Most of us must put our minds to coping with real problems in our daily lives—career, family, health, and so on—yet many of us still find time to enjoy solving puzzles and playing mentally stimulating games. It is precisely this enjoyment factor that makes it possible to maintain a mental aerobics program over the long haul.

Beginning Exercises

1. ***Warm-up Exercise.*** Take a piece of paper and a pencil and try writing your first name using your non-dominant hand (i.e., left hand if you are right-handed). Now take a second pencil and try writing your first name using both hands at the same time. Now try it with your last name.

2. ***Right-Brain Exercise.*** How many squares are there in the following figure?

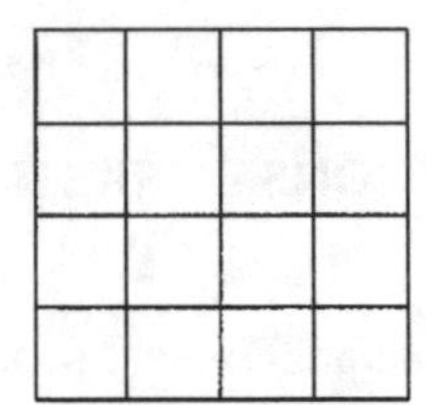

3. ***Right-Brain Exercise.*** Complete the sequence by choosing object A, B, or C:

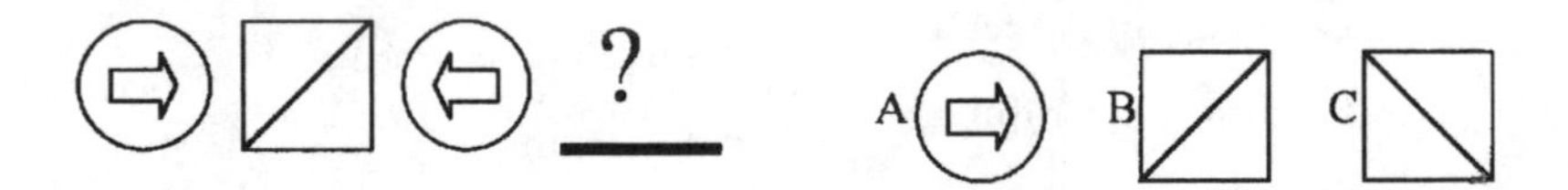

4. ***Right-Brain Exercise.*** Look at the object on the left and then choose the version that matches, A, B, or C.

5. ***Right-Brain Exercise.*** Arrange five toothpicks of your own into the shape of a number five as below. Now try to rearrange them into the number sixteen—without breaking them!

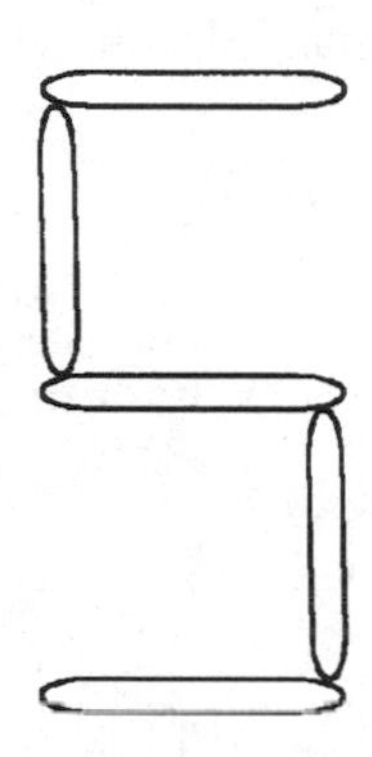

6. ***Left-Brain Exercise.*** The following proverb has had all of the vowels taken out, and the remaining letters broken up into groups of four or three letters each. Replace the vowels and find the proverb:

RLLN GSTN GTH RSN MSS

7. ***Left-Brain Exercise.*** Starting with SOFT, change one letter at a time until you have the word LENS. Each change must be a proper word.

SOFT
• • • •
• • • •
• • • •
LENS

8. ***Left-Brain Exercise.*** A water lily doubles its size every day in a round pond, and after 20 days, the lily will completely cover the pond. How many days will it take to cover half the pond?

9. ***Left-Brain Exercise.*** Which is the odd one out:

CAT MONKEY WHALE MOUSE SHARK

10. ***Left-Brain Exercise.*** What number ends this sequence?

36 25 16 9 _

11. ***Left-Brain Exercise.*** Which letter or number is the odd one in each rectangle?

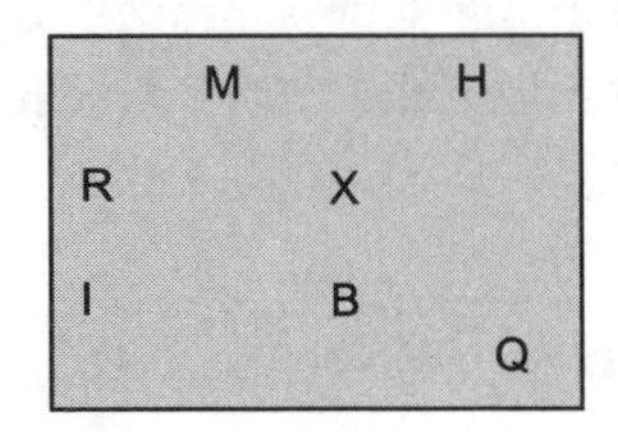

23 21
18 39
16 46 28

12. ***Whole-Brain Exercise (both hemispheres).*** A woman marries 11 men in the space of 10 years. She divorces none of them, none of them die, and she has not committed any crime. How is this possible?

13. ***Whole-Brain Exercise.*** You need to get a pair of matching socks from your drawer but the room is pitch black. You know there are 10 blue socks and 10 brown socks in the drawer. How many socks do you need to remove to be sure you have a pair of matching socks?

14. ***Whole-Brain Exercise.*** Hans is standing behind Gerrie and at the same time Gerrie is standing behind Hans. How can this be?

Answers to Beginning Exercises

1. ***Warm-up Exercise.*** No right answer.

2. ***Right-Brain Exercise.*** The total number of squares is 26 (don't forget all the combinations of squares within squares).

3. ***Right-Brain Exercise.*** C.

4. ***Right-Brain Exercise.*** B.

5. ***Right-Brain Exercise.***

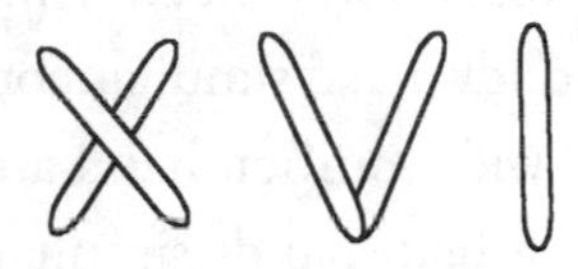

6. ***Left-Brain Exercise.*** A rolling stone gathers no moss.

7. ***Left-Brain Exercise.*** SOFT, LOFT, LEFT, LENT, LENS.

8. ***Left-Brain Exercise.*** It will take 20 days to cover the entire pond, so half the pond will be covered in 19 days.

9. ***Left-Brain Exercise.*** Shark; all of the others are mammals.

10. ***Left-Brain Exercise.*** The answer is 4 (the square of 2) since the numbers are squares of the sequence 6, 5, 4, 3, and 2.

11. ***Left-Brain Exercise.*** In the first box, the letter *I* is the only vowel. In the second box, the number 23 is the only one that cannot be divided by 2 or 3.

12. ***Whole-Brain Exercise.*** She is a minister.

13. ***Whole-Brain Exercise.*** Three socks. If your first sock is blue and your second sock is brown, the third will have to make a pair with one of the first two.

14. ***Whole-Brain Exercise.*** Hans and Gerrie are standing with their backs to each other.

If you are having fun without frustration at this exercise level, you might want to check out the latest websites and other resources for puzzles and brain-teasers with similar levels of difficulties.

Advancing to the next level gets a bit more challenging. Sample a few to see if they are challenging yet fun.

Intermediate Exercises

1. ***Warm-up Exercise.*** You'll need a piece of paper and two pencils again for a more advanced simultaneous writing exercise. Try writing your first name with your left hand and your last name with your right hand, but use both hands simultaneously. After you get the hang of it, reverse the task: write your last name with your left hand and your first name with your right hand, but do it simultaneously.

2. ***Right-Brain Exercise.*** The following 10 circles are arranged in a triangle. See if you can turn the triangle upside down by moving just 3 circles.

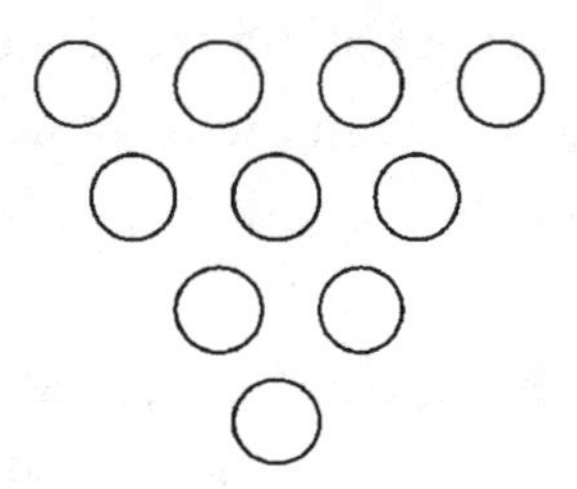

3. ***Right-Brain Exercise.*** Without lifting your pencil from the paper, draw four straight connected lines that go through all nine dots, but through each dot only once. After you have tried two different ways, ask yourself what restrictions you have set for yourself in solving this problem.

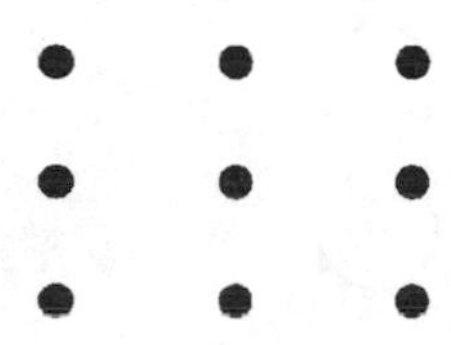

4. ***Right-Brain Exercise.*** These cubes build from the bottom layer up. Figure out the total number of cubes in the figure below, including the number of hidden cubes.

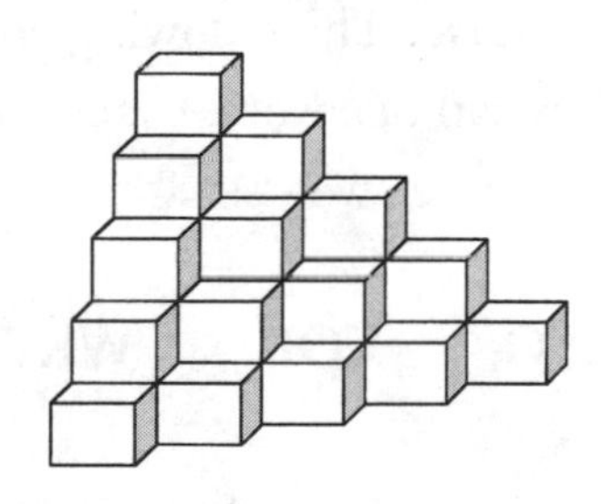

5. ***Right-Brain Exercise.*** Look at the object on the left and then choose the rotated version, A, B, or C.

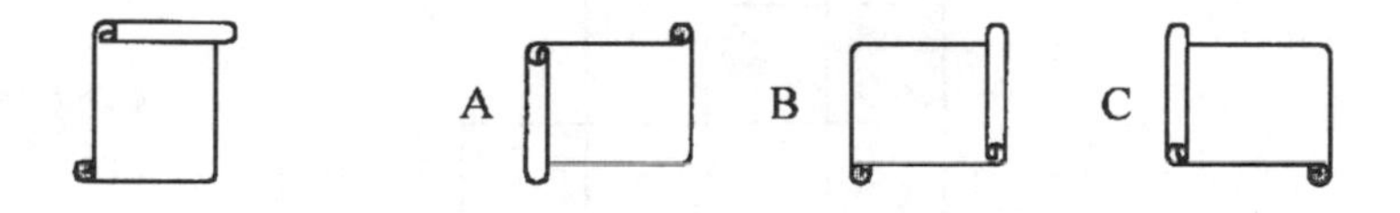

6. ***Right-Brain Exercise.*** Below are six circles. Try to move just one circle to form two rows, each with four circles.

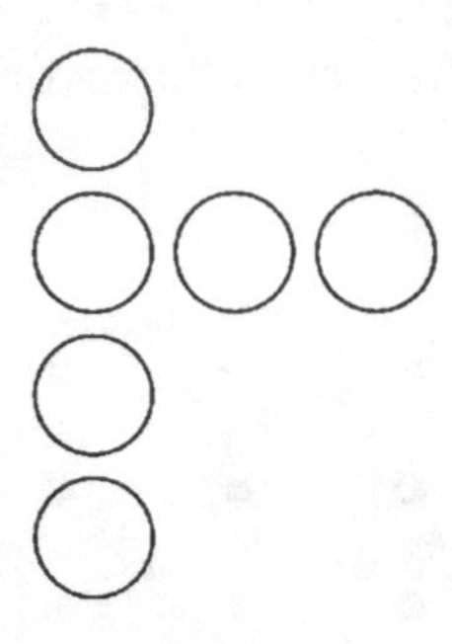

7. ***Left-Brain Exercise.*** Which four colors have been mixed up below?

BYLV GUEE RLLO IWEO ELEN T

8. ***Left-Brain Exercise.*** The following combinations of letters are unusual, but each is part of a word, exactly as they appear in that word. Try to discover the three words:

XYG XOP WKW

9. ***Left-Brain Exercise.*** Using the letters EEEENNNNPPSS, complete the following grid with four words. The words in the grid read the same across as down.

O			
	A		
		D	
			T

10. ***Left-Brain Exercise.*** Can you think of a word that starts with BR, and when you add the letter E to that word, the new word sounds the same as the first?

11. ***Left-Brain Exercise.*** See if you can find the hidden countries below without using any reference material. The letter denotes the country's first letter and the number indicates the number of letters in the country. For example, B6 could be Brazil.

U7, T6, A9, M10, F4, V7

12. ***Left-Brain Exercise.*** Which of the following words is the odd one out?

HAWAH OHADI HATU ADIROLF AIGREOG NOGERO TNOMREV

13. ***Whole-Brain Exercise.*** Shirley has idiosyncratic tastes. She loves weeds but despises flowers. She adores confetti but hates party decorations. She likes feet but dislikes hands. Based on her preference pattern, would she prefer sitting or standing?

14. ***Whole-Brain Exercise.*** Two policemen are patrolling a one-way street looking for drivers who are violating local traffic laws. They see a limo driver going the wrong way down the street, but the policemen do nothing. How would you explain this?

Answers to Intermediate Exercises

1. *Warm-up Exercise.* No right answer.
2. *Right-Brain Exercise.*

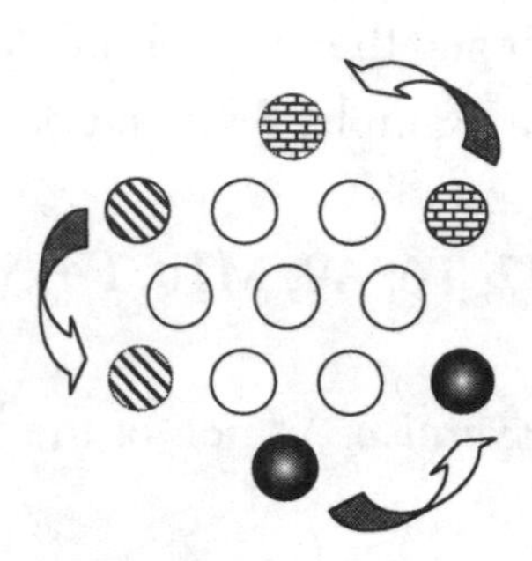

3. *Right-Brain Exercise.* Letting go of our spatial mental assumptions allows us to solve the dot-connecting puzzle.

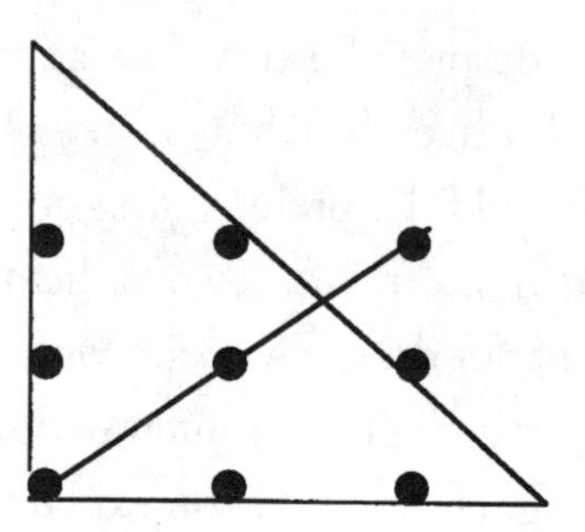

4. *Right-Brain Exercise.* Your total should add up to 35 cubes: 15 cubes are showing and 20 are hidden.
5. *Right-Brain Exercise.* C.

6. ***Right-Brain Exercise.*** Place the left-hand circle under the middle one as shown below.

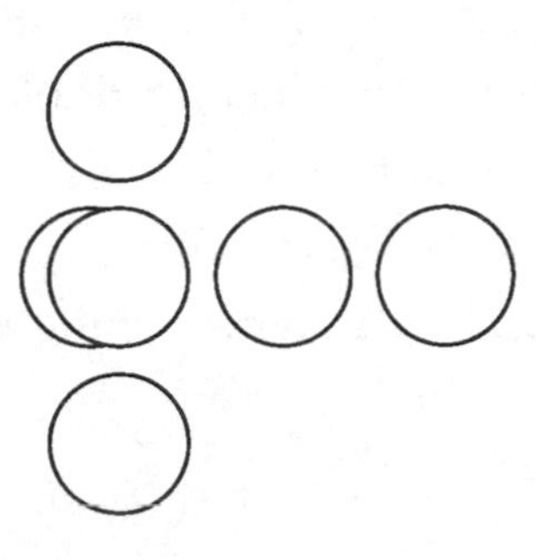

7. ***Left-Brain Exercise.*** Blue, yellow, green, and violet.

8. ***Left-Brain Exercise.*** Oxygen; saxophone; awkward.

9. ***Left-Brain Exercise.*** The words are open, pane, ends, nest.

10. ***Left-Brain Exercise.*** Braking becomes breaking when you add the letter *E*.

11. ***Left-Brain Exercise.*** Uruguay, Taiwan, Argentina, Mozambique, Fiji, Vietnam

12. ***Left-Brain Exercise.*** AIGREOG. All the others spell one of the United States backward. AIGREOG is not the reverse of GEORGIA.

13. ***Whole-Brain Exercise.*** Sitting. She only likes words that contain double letters.

14. ***Whole Brain Exercise.*** Because the limo driver was walking rather than driving, no traffic laws were broken.

If you are not yet mentally exhausted (I know I am), then you are at the top of your mental aerobics game and may wish to move on to the following advanced exercises.

Advanced Exercises

1. ***Warm-up Exercise.*** You'll need a pencil and piece of paper for the warm-up. Take the piece of paper, hold it against your forehead and write your first name. View the results. Try writing your last name with the paper against your forehead. Now try this exercise again while standing in front of the mirror.

2. ***Right-Brain Exercise.*** Which of the following shapes is different from the rest?

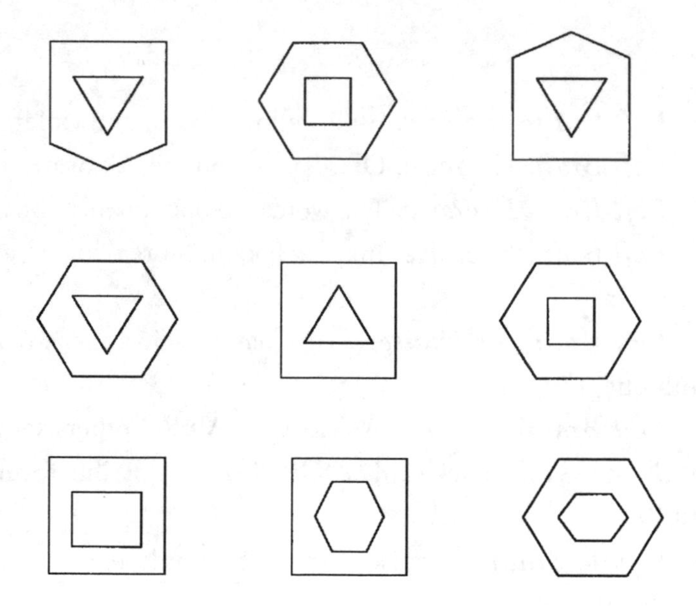

3. ***Right-Brain Exercise.*** Identify the square that completes the sequence, A, B, C, D, E, or F.

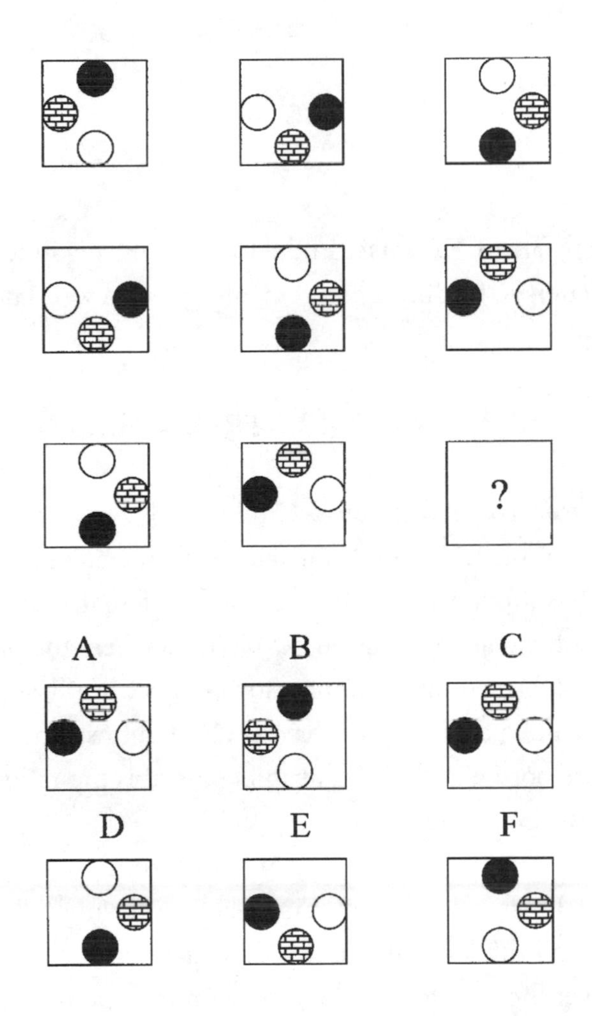

4. ***Left-Brain Exercise.*** Can you unscramble the letters below to find four cheeses?

CCEDHBHEAEMRSDIHEDIARRE

5. ***Left-Brain Exercise.*** Can you circle exactly four of these numbers such that the total is twelve?

1	6	1
6	1	6
1	6	1
6	1	6

6. ***Left-Brain Exercise.*** In the following string of letters, cross out nine letters so that the letters remaining spell a well-known appliance.

RNEIFNRIEGLEETRATTOERSR

7. ***Whole-Brain Exercise.*** China has been grappling with a population problem for some time. For many social and cultural reasons, families strongly prefer male children to female children. Consider a hypothetical city somewhere in China where the practice has arisen that every family continues to procreate until a son is produced, at which point they stop having children. Assuming that boys and girls are born with equal probability, what is the ratio of boys to girls after 100 generations?

8. ***Whole-Brain Exercise.*** You return from work and discover that your television is on. Not remembering having left it on, you turn it off and think nothing of it. A few days later, the same thing occurs. Over the next few weeks, it happens several more times and then stops. Deciding this case did not warrant calling in Mulder and Scully, or even Robert Stack, you forget it. Now, several months later, it has begun again. The baffling facts of the case are as follows:

- **You have never observed it occurring when you are home.**

- **All the doors and windows in your house are locked when you leave.**
- **There is no sign of trespass when you return.**
- **No one is at home while you are gone except your pet goldfish, Emma, who really is more of a radio fish.**
- **Your remote control's batteries have been dead for some time.**
- **When the TV is found on, it is a seemingly random channel: news, soaps, static, or *Baywatch*.**
- **The television room is on the top floor, and there are no houses, buildings, or other structures within line of sight of any window.**
- **No other appliance in the house displays this behavior.**
- **There seems to be nothing odd about your electrical system—no surges or spikes.**
- **The TV is still under warranty and passed its most recent inspection by a trained technician.**

Is your TV possessed by an unearthly couch potato or can you think of a more mundane explanation?

9. ***Whole-Brain Exercise.*** What is the missing letter?

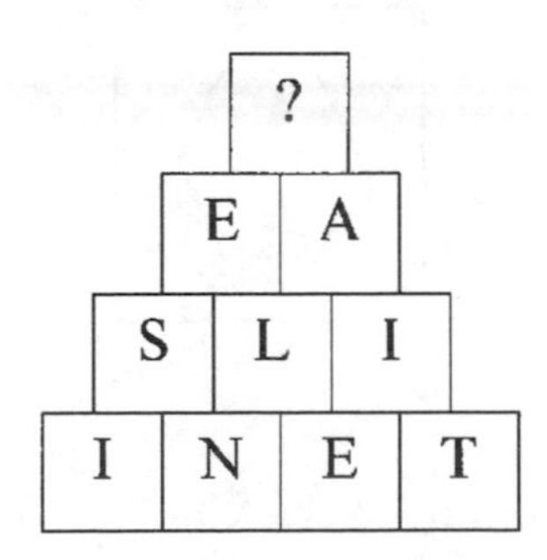

10. ***Whole-Brain Exercise.*** Three men, Alan, Brian, and Charles, and their respective wives, Alice, Betty, and Cathy, were hunting in Africa, when they came across a large river. Luckily there was one boat, but it could only carry two people at the same time. Due to bitter jealousy, no woman could be left with another man unless her husband was present. How did they manage to cross the river?

11. ***Whole-Brain Exercise.*** There is a closed room with a light in it. Outside, there are three light switches. You can flick any of the switches any number of times, but only one at a time. You can only open the door and go into the room once. You know that the light is initially off. How can you determine which light switch operates the light?

Answers to Advanced Exercises

1. ***Warm-up Exercise.*** No right answer.

2. ***Right-Brain Exercise.*** The only piece with more sides on the inner shape than the outer one.

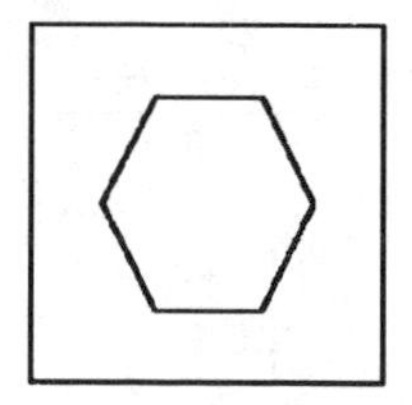

3. ***Right-Brain Exercise.*** B. Moving from left to right and top to bottom, the black and white circles move clockwise each step and the brick circle moves counterclockwise.

4. ***Left-Brain Exercise.*** Edam, cheddar, brie, cheshire.

5. ***Left-Brain Exercise.*** Turn the grid upside down.

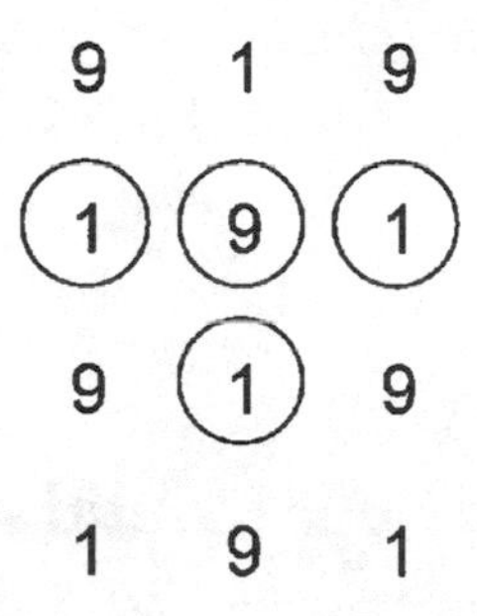

6. ***Left-Brain Exercise.*** If you cross out the letters NINE LETTERS, you spell REFRIGERATOR.

7. ***Whole-Brain Exercise.*** The ratio is 50–50. Intuitively, it may feel that the families are adopting a strategy favoring producing a son, but this is incorrect. Each family's expected number of sons is one, by the definition of the strategy. But the expected number of daughters doesn't drop.

8. ***Whole-Brain Exercise.*** Just like in the *X-Files*, the answer is that your TV is possessed by an unearthly couch potato: the sun. Your TV room's windows face west, and during certain times of the year, the sun's refracted rays come streaming in and strike your television's remote control sensor. Since the remote operates on infrared light, certain frequencies of light emitted by the sun are interpreted by the sensor as commands to turn on or off, or change the channel. When you are home, you pull the shades, to keep out the glare of the setting sun.

9. ***Whole-Brain Exercise.*** N. The pyramid spells out the name AL EINSTEIN.

10. ***Whole-Brain Exercise.*** Alan and Alice cross and Alan returns. Betty and Cathy cross and Alice returns. Brian and Charles

cross and Brian and Betty return. Alan and Brian cross and Cathy returns. Alice and Betty cross and Charles returns. Charles and Cathy make the final crossing.

11. Whole-Brain Exercise. Leave switch 1 alone. Flick switch 2 on for an hour, then flick it back. Flick switch 3. Now look. If the light is on, then the switch is 3. If the light bulb is warm, then it is switch 2. If not, then it is switch 1.

Building Your Mental Aerobics Program

Just as physical activity can keep your body strong, mental activity can keep your mind sharp and agile. You can continue to challenge yourself by using a variety of approaches. You might consider exploring a new hobby, learning a foreign language, or perhaps taking up a musical instrument. Making a change in your leisure reading—perhaps switching from romance novels to biographies or mysteries—could potentially tweak your dendrites.

Whether you were able to complete all of the exercises or only a few, you should have a sense of the difficulty level for mental aerobics exercises that suits you. As you build your skills over time, you may want to advance to a higher level to challenge yourself and keep you stimulated. Chapter 10 will help you fit a program into your weekly schedule, and you can readily expand your repertoire with novel puzzles, games, and brain-teasers from other sources, including magazines, books, and websites.

Chapter Six

Build Your Memory Skills Beyond the Basics

I am always ready to learn although I do not always like being taught.
—WINSTON CHURCHILL

Whether it's riding a bicycle, using a typewriter, or ironing a shirt, we take for granted most of the skills we learn throughout our lives. Yet, for each of these routine activities we had to build slowly upon basic steps to master a more complex activity. The same holds true for memory skills. Just as we systematically learn how to master everyday basic tasks—driving a car, using a hand-held organizer—we can systematically learn memory techniques and incorporate them into our daily routines.

In Chapter 3, we saw how *LOOK, SNAP, CONNECT*, the basic building blocks for my memory training program, could improve our memory performance quickly. If you have begun to utilize these skills with ease, you are ready to address the next level of memory skills training.

Organization

A professor of mine once commented on the superb skills of a very accomplished scientist at UCLA. She attributed much of his accomplishments to his being "extremely organized." That statement made an impression on me—she didn't describe him as brilliant, creative, insightful, or scholarly. She merely said he was organized.

Rhonda C. and her husband Ken had postponed having children well into their thirties because their advertising careers were in high gear. After Ben and Nikki were born they really didn't need two incomes, so Rhonda stopped working. Now, in her mid-forties, she was up to her ears in managing the house; driving carpools; attending school functions, soccer practices, ballet lessons, and her husband's social events; and taking care of what seemed like a million other little things. Rhonda never saw her mother's broken hip coming.

Her 78-year-old mother had always been fiercely self-sufficient. She was an avid tennis player, member of a bridge club, and she loved to travel. Rhonda had been seeing her mother once a month or so, even though they lived less than ten miles apart. Now there were doctor appointments, medications, groceries, and a list of other errands and personal requirements that mother needed done and only one person for the job: Rhonda. Mother refused to have "some stranger" come to straighten her house and do laundry, so Rhonda did it. Mother wouldn't dare have "pre-cooked junk" meals delivered, even from good restaurants, so Rhonda cooked for her mother before going home to cook for her own family.

Rhonda began waking two or three times a night feeling anxious—had she forgotten to do something? Left something unfinished or not done well enough? She constantly felt guilty about spending too little time with the kids, Ken, and even her mother. She hadn't had time to go to the gym or take a run for weeks, and then her memory started to go. First she forgot to pick Ben up from basketball clinic—twice. Then little Nikki was inconsolable when Rhonda "mixed up the dates" and didn't show up to do an art project with her second-grade class. Ken realized Rhonda might be having a real problem when he came in with an important out-of-town client for a home-cooked dinner, but Rhonda had "spaced out" and ordered pizzas as a treat for the kids.

Rhonda was exhausted. She lost weight and grew restless. She was constantly bickering with everyone, especially her mother. She felt terrible but she couldn't stop—there were always three or four more things she had to do first. Rhonda became depressed and her memory got even worse.

Ken convinced her to see a therapist they'd heard of. Once there, she broke down and sobbed that she was losing it. She used to run a successful advertising agency and now she couldn't even manage a household! She felt overwhelmed, stressed out, and disorganized. She could hardly remember all the errands and tasks she had to do each day.

The therapist said she appeared to be under severe stress and was definitely depressed, and often when people are depressed, it affects their memory ability. Under some circumstances, he might prescribe an antidepressant, but he was struck by how much of her problems began when her mother broke her hip and Rhonda suddenly had to add parent care to her many tasks. Rhonda balked: at work she used to have hundreds of tasks and a dozen employees vying for her time. Every minute of every day was scheduled, and she never had memory problems then. The therapist said that might just be the answer: when Rhonda was working, she adhered to a well-organized, scheduled agenda. With her new "job," she needed to create a similar, effective system to organize her schedule.

Rhonda knew he was right. Organizing her new life was something she had neglected, long before her mother's accident. She went home and began listing her many daily tasks and chores, and then sorted them by priority and geography. She bought an appointment organizer like the one she had used in her agency days and began scheduling her upcoming week. Rhonda felt empowered, and although she was relying on her appointment book as a memory tool, the process of going back to her familiar organizational strategies made it easier for her to remember her engagements without actually having to refer to the book itself.

Rhonda was able to see her mother back to health and never did need those antidepressants.

Organization is essentially the process of systematically arranging information according to structures, patterns, and groupings. Learning to organize daily activities effectively can make the difference between success, mediocrity, or out-right failure. For maximum recall performance, organizing information according to obvious patterns facilitates quick memory storage and retrieval.

One of the more effective organizational memory skills involves *chunking*—basically, dividing a large group of random items into separate chunks with a common characteristic. Attempting to remember six random items at the market is going to be more difficult than trying to remember three cereals and three dairy products.

Consider the pile of common grocery items below:

Many of us would find them easier to remember if instead of seeing a random assortment, we split them up into two groups, or chunks: in this case, three fruits and three meats.

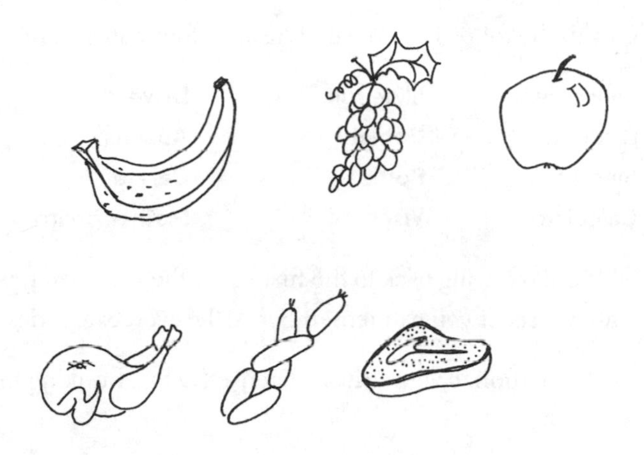

Four Organization Exercises

1. Group the following twelve items into three categories:

Hammer	**Carrot**	**Golf club**
Nail	**Hoop**	**Scalpel**
Cucumber	**Romaine**	**Cabbage**
Base	**Stopwatch**	**Bolt**

   Without looking at the above list, try to remember the items in each of your categories.

2. Dividing information into smaller groups or clusters is another form of chunking. For example, it is easier to remember three chunks of two- or three-digit numbers than an entire seven-digit phone number: 82-51-291 instead of 8251291. Some boomers may remember when phone numbers included a word like "Webster" representing the first three numbers. Look up the phone number

for your public library. Read it once, cover it, and say it out loud. Difficult? For many of us it will be. Now look at it again and cluster it into three smaller number groups. Now say it aloud. You may find it easier to remember.

3. Group the following twelve items into four categories:

Sunscreen	**Slide rule**	**Drive**
Diskette	**Goblet**	**Abacus**
Shutter	**Bottle**	**Carafe**
Calculator	**Visor**	**Mother board**

Without referring back to the first list or the list in exercise 1 above, see if you can remember ALL seven categories.

4. Try to memorize your driver's license ID by chunking the numbers.

Peg Method for Remembering Numerical Sequences

With the advent of cell phones, fax machines, e-mail, and pagers, many baby boomers find themselves suffering from numerical technology overload, or too many darn numbers to keep straight. One might wonder: "Who needs to remember numbers when there's so many electronic phone book gadgets available? Why, I have one right here in my bag!" But what if your bag *isn't* right here, and you must call your boss's cell phone immediately? With a simple *Peg Method*, you can know your boss's cell phone number and never, ever forget it.

The Peg Method was developed as a system for remembering phone numbers, addresses, or numerical sequences by visualizing objects as opposed to rote memorization of the numbers themselves.

Just as a peg is something that pins down or fastens things, this technique helps us to systematically pin down or fasten bits of information. Pegging builds on the linking skills we learned in *LOOK, SNAP, CONNECT* by providing a way to remember items in any order one chooses, as opposed to linking, where we are limited to remembering information in its original sequence.

Although it requires effort, the Peg Method will forever remove uncertainty about remembering numbers—any sequence of numbers—whether it's phone numbers, combinations, passwords, or social security numbers. You will be able to punch in your credit card number *and* its expiration date and never remove it from your wallet. Of course, pegging can prove to be challenging, so those satisfied with the linking method they mastered in *LOOK, SNAP, CONNECT* may choose to jump to the next section.

To use the Peg Method, you will need to commit to memory ten specific, simple visual images—one for each of the ten numerical digits. Begin by assigning each of the numbers one consonant letter of the alphabet that reminds you of that number. For example, I use the letter *T* to represent the number *1* because it has *one* downstroke. I then use this letter to begin a word that invokes a visible image, in this example, a tie, and this word *tie*, then, serves as my peg for the number one.

In Table 6.1, I provide a sample peg word for each of the ten digits. If you like, you can learn these words or make up your own peg words based on your first association to the consonant sounds, and write them in your notebook.

Table 6.1

Number	Word	Consonant	Memory Aid
1	Tie	T	Letter with 1 downstroke.
2	Nun	N	Letter with 2 downstrokes
3	Mummy	M	Letter with 3 downstrokes
4	Raisin	R	Last sound of the word "four."
5	Leg	L	Roman numeral for 50 is *L*.
6	Jet	J	Letter *J* turned around looks like a *6*.
7	Kite	K	*K* looks like two 7s rotated and glued together.
8	Fork	F	Lowercase cursive *f* looks like the number *8*.
9	Pot	P	Letter *P* turned around looks like *9*.
0	Zebra	Z	First letter in the word "zero."

In addition to being readily visualized, each of these words leads to distinct images. A raisin is hard to confuse with a mummy, and a pot can be readily distinguished from a fork. Effective peg words also tend to bring to mind many varied details, which lead to more memorable visual images, or snaps.

Using pegs and the link method together can also be very effective. Here is how you might remember your wife's social security number using the above peg words.

SSN: 557-16-8043

Peg word sequence: Leg
Leg
Kite
Tie
Jet
Fork
Zebra
Raisin
Mummy

Storyline-linking peg words: Imagine your wife sitting and crossing her *legs* (55). (You could never forget this is her social security number since she has such great legs!) She loves being outdoors so she has decided to fly a *kite* (7) while sitting with her legs crossed. Let your eye follow the kite string upward to where it connects to the kite and you see a *tie* (1) waving in the wind as its tail. But the tie-tail suddenly gets sucked into the engine of a passing *jet* (6). You notice that the pilot is actually a humanoid *fork* (8), and a *zebra*, (0), dressed as a flight attendant, enters the cockpit to serve the fork a large *raisin* (4). As the fork tries to figure out how to eat the raisin, a *mummy* (3) dressed as the co-pilot enters the cockpit.

Admittedly, this is a rather bizarre sequence of events and would not occur often on a major commercial airline. But keep in mind, the more bizarre, unusual, and vivid your peg images, the easier they are to commit to memory. Also, these associations were the first ones that came to my mind, and oftentimes first associations stick best.

Review and memorize the peg word list in Table 6.2. Then, using your notebook, practice the Peg Method with the exercises in the Peg Method Exercises box.

Table 6.2

SAMPLE PEGS	
DIGIT	WORD
1	Tie
2	Nun
3	Mummy
4	Raisin
5	Leg
6	Jet
7	Kite
8	Fork
9	Pot
0	Zebra

Peg Method Exercises

1. B*irth dates*. Write down a friend's birth date, for example: 09/21/54. Find the corresponding peg words and then create a story—as zany and vivid as possible. Now use the same method to commit to memory two more birth dates, those of a relative and a work associate.
2. *Phone numbers*. Use the Peg Method to remember the phone numbers you don't already know of the following businesses: a favorite local restaurant, a good plumber,

and a drugstore. For extra credit (Type A personalities): learn their fax numbers too.

3. *Credit card numbers*. Do your two most-used cards. Don't forget the expiration dates.

Remembering Names and Faces

One of the most frustrating things about the memory changes associated with aging is the increasing difficulty we have remembering names. We may recognize a person's face but be unable to recall the person's name. Even as young adults, the major reason many of us forget names, sometimes only seconds after people have been introduced to us, is that oftentimes we are not fully listening. Fortunately, there are many easy-to-learn strategies for remembering names.

Most strategies for remembering the name that goes with a face use the three basic skills we learned in Chapter 3: *LOOK, SNAP, CONNECT*. First, make sure you consciously *listen to* and *observe* the name (*LOOK*). Then, *SNAP* a visual image of the name and the face. Finally, *CONNECT* the name-snap with the face-snap. This systematic approach of linking a name to a face has been highly successful.

Other effective techniques include repeating the person's name during your initial conversation, as well as commenting on how the person reminds you of someone else you know of the same name. Other mnemonics work well, also. I mentioned my challenge every school year, learning the names of my kids' new classmates and their parents. My son's first-grade class was particularly challenging (I'm not sure why). However, one little girl was named Ashley and her mother was Laura. Laura; Ashley. Since my wife has an expensive yet seldom-indulged penchant for bedding by a company of that name, I found the names of this child and her mother very easy to remember.

If a person has a complicated or unfamiliar name, you might ask them to spell it for you, or sometimes just visualizing an image of the name spelled out will facilitate memory storage. Using their name when saying good-bye will certainly help secure it into your memory banks.

All names can be placed into two groups: those that have a meaning and invoke a visual image, and those that don't. Names like "Carpenter," "Katz," "House," "Bishop," "Siegel," "White," or "Silver" all have a meaning and can readily bring to mind an image. Of course, I am thinking of a seagull when I meet Mrs. Siegel.

Often names may have no immediate meaning but can still bring an image to mind. For example, "Bill" could be represented by the image of a dollar bill. The name "Washington" might conjure up the famous monument.

Any name that has no immediate meaning, like "Shapiro," "O'Malley," or "Amaducci," may require additional mental effort to remember. However, the names or the syllables and sounds within them can be associated to a substitute name or sound that does have a meaning. By linking these substitute words together, you can create a visual image that works. Sometimes we can break a name into syllables that contain meanings, and then *link* them afterward. For example, the name "George Waters" could be remembered through an image of a *gorge* with a stream of *water* rushing into it. The word or syllable substitutes do not need to be exact. "Frank Kaufman" could be a *frank*furter being eaten by a *coughing man*. "Gene Phillips" could be a Phillips screwdriver wearing a pair of tight jeans. Of course, some scientists among us might prefer visualizing a double-helix gene downing a double-vodka screwdriver.

Finally, we need to *CONNECT* the name to the face. The

approach here is to look at the person's face and search for a distinguishing feature, whether it is a small nose, large ears, unusual hairdo, or deep wrinkles. Just pick the first outstanding feature you notice and link it to the name. For example, if Mrs. Stockton has a round face, think of a balloon falling on a *ton* of *stock* certificates.

To create a snap for remembering Mr. Bender's name, you may see him bending forward, as below.

Naturally, others might choose to picture him disheveled from a "bender" the evening before, when he had one too many.

Often, the first thing that strikes us about people is not visual but instead relates to their personality. Mr. Porter has a great sense of humor and a silly laugh, so we might visualize him as a *porter* carrying our suitcases, dressed as a circus clown. Again, the images and substitute words need not be perfect. The process of thinking up the images and making the connections, or links, will fix them into memory. Of course, practice makes perfect, and it can be found in the Name and Face Exercises box.

Name and Face Exercises

1. For the following first names, think of a visual image:

NAME	VISUAL IMAGE
Stewart	
Cheryl	

2. Write down the last name of two people you know and create a mental snapshot that represents the name:

PERSON'S NAME	VISUAL IMAGE

3. For the above people, list the first distinguishing feature that comes to mind:

PERSON	FEATURE NAME/FACE IMAGE

The Roman Room Method

To apply the Roman Room method, originally developed by ancient Roman orators to help them remember long speeches, first visualize a familiar room. Then place each item to be remembered in a specific location there. You can then retrieve the information when taking a mental walk around the room.

The method can be useful for speeches, lectures, or lists. You might imagine your living room, bedroom, or office as your "Roman room." I can imagine myself in my office where I see from left to right the *computer, phone, bookshelf,* and *couch*. I need to remember the following errands after work: *hardware store, carwash, cleaners,* and *market*. I then visualize a wrench on top of the desk, a wet car on the phone, folded shirts in the bookshelf, and a shopping bag on the couch.

A variation uses a familiar route rather than a familiar room. On your commute to the office, you pass the following landmarks each day: water tower, bridge, gas station, and post office. You need to thank several people in your Academy Award speech so you think of Steven Spielberg sitting atop a water tower, Meryl Streep walking across the bridge, Sylvester Stallone pumping gas at the station, and Jerry Lewis working at the post office.

Building Memory Skills: A Quick Review

1. *Organization*. Look for systematic patterns and groupings to facilitate learning and recall.
2. *Peg Method for Remembering Numerical Sequences*. Commit to memory a specific visual "peg" for each of the ten digits; then use the link method to create a story for remembering numerical sequences.
3. *Remembering Names and Faces*. Make sure you consciously listen and observe the name (*LOOK*), then *SNAP* a visual image of the name and the face, and finally *CONNECT* the name-snap to the face-snap.
    - **Observe distinguishing features in the person's face.**
    - **Repeat the name in conversation and when saying good-bye.**

- **Look for personal meaning in the name.**
- **Ask them to spell their unusual name.**

4. *Roman Room Method.* Pick a familiar room or route and, in your mind, place the items to remember at key points or landmarks.

Chapter Seven

Start Your Healthy Brain Diet Now

I drive way too fast to worry about cholesterol.
—Steven Wright

Most of us realize by now that the quality and quantity of the foods we eat affect our bodies and our physical health. Less widely understood is the critical impact our nutritional habits have on our *brain* health—particularly our memory performance and risk for developing Alzheimer's disease.

Just as unhealthy diets can lead to physical ailments like diabetes, heart disease, and obesity, those same T-bone steaks, curly fries, and ice cream sundaes can negatively, and sometimes irreversibly, damage our brain fitness—although the effects may take decades to appear.

Convincing scientific evidence indicates that long-term, healthy dietary habits may prevent future brain aging and memory decline and help protect our brains from developing Alzheimer's disease symptoms. For many people, even a few weeks of healthy, low-fat eating can produce immediate benefits such as increased alertness and greater energy.

The sooner we start our healthy brain diets, the sooner we begin to reap the benefits. Chapter 7 provides the components of a safe mental-fitness diet to keep your brain young and protect it from Alzheimer's disease.

Calories Count to Healthier Brains

By the time we reach middle age, many of us tend to carry around extra body weight. Whether it's simply an additional five or ten pounds, or true obesity, excess body fat increases our risk for illnesses like diabetes and high blood pressure. These illnesses increase our risk for small strokes in the brain, which can lead to memory decline and dementia and even Alzheimer's disease.

Among the most effective and widely accepted ways to drop weight and avoid these problems is limiting daily caloric intake while increasing physical activity—a practice at the heart of almost all traditional weight-loss programs. Any reputable book or program on dieting or weight reduction, no matter how miraculous its "breakthrough" methods may be, comes down to these tenets at some point. Who of us struggling against love handles wouldn't want to "Lose weight while sleeping!" or try a month on "The more you eat the more you lose!" diet? Alas, the FDA has not yet approved the "Magical Fat Melters" and their ilk. Following a few practical tips has helped me maintain a reasonable diet and watch my calories (see box).

Dr. Roy Walford at UCLA is among the gerontologists conducting animal studies showing that lifelong calorie restriction dramatically prolongs life expectancy as well as maintains optimal brain fitness. Recently, calorie-restricted rats were found to have 25 percent better functioning of brain receptors involved in memory compared with animals on unrestricted diets.

Dr. Walford told me that he was so convinced by his animal

Practical Tips for Keeping Tabs on Your Calories

- **Drink plenty of water, at least six glasses a day.**
- **Plan your meals in advance. Don't wait until you're so hungry that you'll eat junk food or overeat.**
- **Keep portions low, avoid large meals, and eat healthy between-meal snacks to avoid a sense of deprivation.**
- **Use spices, herbs, garlic, salsa, and other healthy taste enhancers.**
- **When dining out, try splitting an entrée with a friend.**
- **Cook or order small portions to avoid overeating.**
- **Limit nighttime snacking. Brush your teeth an hour or two before bedtime as a reminder.**
- **Consider substituting the following lower-calorie food choices for their higher-calorie counterparts:**

YES	NO
Fish or poultry breast ←	→ Red meat
Non-fat frozen yogurt ←	→ Ice cream
Fresh fruit ←	→ Sweetened canned fruit
Low-fat or skim milk ←	→ Whole milk
Unbuttered popcorn ←	→ Buttered popcorn

studies that he had begun restricting his own calories by fasting at least one day each week in the hopes of maximizing his longevity. Although I do not recommend caloric restriction to maximize memory ability to my patients, I do think that for people who are overweight or obese, watching their calories can decrease their risk of developing weight-related illnesses and help slow down brain aging.

Yo-Yo Syndrome

Another joy of aging is our bodies' gradual loss of its automatic ability to regulate appetite and maintain a constant body weight. In a recent study, both younger and older men were asked to eat approximately 1,000 calories above their normal daily intake. After several weeks, this high-calorie addition to their diet ended. The younger group automatically ate less and dropped back to their normal body weight without trying to reduce. The older group kept the weight on. This older group of men was similarly unable to bounce back from undereating. After three weeks on a reduced diet—about 800 calories less than usual—all were asked to return to eating normally. The older group tended to stay at their lower weight level, while the younger group gained back what they lost in the six-week follow-up.

One of the greatest problems I see with most diets is that people get into a "yo-yo" syndrome of going up and down in their body weight. They begin a crash diet, feel starved and deprived, and then go off the diet with a giant binge. Experts agree that in the long run, such yo-yo diets tend to backfire and actually lead to increased body weight.

Researchers put healthy laboratory rats on a yo-yo diet and found they developed 3 to 4 percent more body fat than animals kept on a steady caloric intake. Weight cycling, or the pattern of gaining, losing, and regaining weight, often leads to greater fat accumulation over time. The evidence from animal studies and the harmful physical and mental effects of obesity has convinced many experts that by sensibly watching our caloric intake we may help protect our brains.

Georgette B., a 46-year-old teacher, first sought help for her memory problems two years ago. She was already being treated for high blood cholesterol, and she was concerned about her family's history of Alzheimer's disease. During the first five minutes of her visit, it was clear that memory complaints were not her only problem. In describing herself, she said she was always the first of her friends to try the latest fad diet. She would typically begin a new diet regimen, maybe lose a few pounds in the first week or two, and then quickly grow discouraged as her weight loss tapered off. Georgette would then abandon the diet and almost ritualistically plunge into a post-diet binge to make up for her weeks of hunger and deprivation. Experiencing a sense of failure and defeat, she invariably became depressed until finally finding and latching onto the next miracle weight-loss program. During her bouts of depression, Georgette's memory difficulties became markedly worse. It was clear that if I were going to help Georgette with her memory problems, her eating disorder would need to be addressed as well.

Her approach to dieting had been like a physical and emotional roller coaster. She understood that what she ate had an effect on her waistline, but she never imagined that what she ate could have an impact on her current memory and future ability to think.

I referred Georgette to a nutritionist, who helped design a diet that kept her satisfied. By feeling less deprivation, she was less likely to quit the diet and start binging.

Another issue for Georgette was chronic depression. After treatment with an antidepressant, her memory complaints diminished and she found it easier to stick with her diet. For Georgette, learning that food choices could ultimately affect her brain aging and risk for developing

> Alzheimer's disease helped change her attitude about food and eating. She became more realistic in her dietary goals and began to focus on the type and quality of food choices rather than just calorie counting.

A healthy brain diet is not only about counting calories or losing or gaining weight. It involves learning to make simple, consistent food choices that incorporate common sense and some easily learned tips on what to look for and what to avoid.

Good Fats, Bad Fats, and the Men and Women Who Love Them

Dr. Hugh Hendrie of Indiana University compared rates of Alzheimer's disease in African-Americans living in Indianapolis, Indiana, with those of Africans living in the Nigerian city of Ibadan. The Americans developed dementia at a rate nearly three times greater than the Nigerians. The Americans also had higher rates of hypertension, high cholesterol, and diabetes than did the Nigerians, and these illnesses all contribute to the risk for dementia, particularly the vascular form.

Although genetic risk likely plays a role, the people of Ibadan, Nigeria, are mostly poor and unable to afford much more than vegetables to eat. Their diet consists of yams, palm oil, a small amount of fish, and other foods. This diet contrasts dramatically with the typical American diet, which is usually high in animal fats. The researchers judged the Nigerian diet, normally low in protein, fat, and total calories, to be a major contributor to the lower rate of dementia in their population.

Dr. Jim Joseph of the U.S. Department of Agriculture has

described the importance of dietary fats among a Native American tribe in New Mexico with similar genetics to their Mexican counterparts. The New Mexico portion of the tribe lives on "government food," which includes processed flour, cheese, and related high-fat foods. Many are overweight and have early-onset diabetes. By contrast, their relatives in Mexico eat a healthier diet of rice and beans, and despite their identical genetic makeup, obesity and diabetes are unknown in these people. Native Alaskans on the Kenai Peninsula and their genetically identical relatives in Siberia share a similar story. The Native Alaskans are overweight and eat the "white man's diet," while the healthier Siberians eat from the land.

Some recent, popular weight-loss diets advocate eating generous portions of animal proteins and fats, while minimizing or eliminating carbohydrate intake altogether. Despite the effectiveness of these diets in reducing body weight, often due to loss of body fluids associated with carbohydrate restriction, their ability to slow down brain aging is questionable, and these diets may increase the risk for heart disease, diabetes, and cancer.

Doctors agree that a healthier diet option might involve limiting animal fats and increasing whole grains, vegetables, fruits, and dairy products, whose benefits stem partly from their potassium and calcium contents. Concerned about the harmful effects of fat, many Americans and Europeans have in fact been lowering their fat consumption for several years.

Epidemiologists consistently find that eliminating most fats from our diets lowers our risk for Alzheimer's disease, and it is never too early to begin a low-fat diet to keep our brains young and stave off Alzheimer's disease. Dr. Robert Friedland and his associates at Case Western Reserve University recently reported that lower-fat diets in young and middle-aged adults may substantially reduce their risk for Alzheimer's disease decades later. In fact, limiting fat intake appears to have its greatest benefit for people *with* a genetic risk for age-related memory loss or Alzheimer's disease. Dr. Friedland's group found that people with the APOE-4 Alzheimer's risk gene who ate a low-fat diet had a strikingly lower risk for developing Alzheimer's disease compared with their counterparts who ate fatty diets. A similar lowering of Alzheimer's risk was not observed in people without the genetic risk.

Some experts believe that the APOE-4 Alzheimer's risk gene accelerates age-related memory loss through its effect on fat metabolism. Among its many functions, APOE's protein product acts as a transport mechanism, or chaperone, for cholesterol in the blood. High blood cholesterol levels not only increase our risk for heart disease and stroke but also make us more susceptible to Alzheimer's disease. A recent study found that patients using statin cholesterol-lowering drugs had a 70 percent lower risk for developing Alzheimer's disease (Chapter 9).

High blood pressure, one of the most common chronic diseases associated with aging, also increases people's risk for multiple strokes, which can cause severe memory loss. It is widely known that limiting dietary salt helps lower blood pressure. A recent study in the *New England Journal of Medicine* found that people with high blood pressure also benefited from adding several servings of fruits and vegetables and low-fat dairy foods to their diets.

Not all fats are bad and accelerate brain aging. Some fats actually promote brain fitness. Dietary fats come in four forms: cholesterol, saturated, monounsaturated, and polyunsaturated. Omega-6 and omega-3 fats are polyunsaturates. Omega-3 fatty acids, often considered "good fats," come from foods such as fruits, leafy vegetables, nuts, fish, fish oil, and olive oil. We can also get omega-3 fats as capsules or supplements. By contrast, omega-6 fatty acids, often considered "bad fats," usually come from meat and other animal products. Common foods containing these fats include red meat, whole milk, cheese, margarine, mayonnaise, most processed foods, fried foods, and vegetable oils.

Diets high in omega-6, or bad, fats may contribute to chronic brain inflammation, a possible underlying mechanism in Alzheimer's disease and other neurodegenerative disorders. Omega-3, or good, fats help keep brain cell membranes soft and flexible, while bad fats make them more rigid. Omega-3 fatty acids reduce risk for cardiovascular disease and stroke. The American Heart Association recommends at least two servings of fish each week so people can get enough of those good fats.

A Dutch study of approximately thirteen hundred men found that those eating margarine and other foods high in omega-6 fats experienced more cognitive decline than those who had healthier diets. By contrast, foods rich in omega-3 fatty acids, such as olive oil, decrease the risk for cognitive decline. In a recent investigation of older Italians, their use of approximately three tablespoons of olive oil each day was enough to provide protection against memory loss when compared to a control group not using olive oil.

Omega-6 saturated fats appear to impair memory through their effects on the hormone insulin. Laboratory animals that are fed omega-6 fats have increased difficulty learning and getting through mazes. In addition, their brain cells show fewer branches, or dendrites. Eating omega-6 fats also increases risk for insulin resistance—insulin becomes less effective in getting glucose into cells, putting people at greater risk for the memory impairments associated with diabetes. Fortunately, diet-related insulin resistance can be reversed, and controlling diabetes with diet, weight loss, or drugs can improve memory as well as learning ability.

Clearly, a diet rich in omega-3 fatty acids is likely to benefit our brain fitness and overall health, but our bodies are also able to adapt to a limited amount of omega-6 fats. An occasional donut or slice of apple pie won't necessarily wipe your mother's maiden name from your memory stores. Some nutrition experts even suggest maintaining a ratio of one omega-3 fat for every omega-6 fat, rather than attempting to completely eliminate the bad omega-6 fats from our diets.

Table 7.1 lists a number of common foods containing mostly omega-3 or omega-6 fats. A well-planned healthy brain diet will emphasize foods high in omega-3 fats.

Table 7.1

SOME COMMON FOODS CONTAINING GOOD AND BAD FATS	
GOOD FATS (HIGH IN OMEGA-3)	BAD FATS (HIGH IN OMEGA-6)
Anchovies	Bacon
Avocados	Butter
Bluefish	Cheese
Brazil nuts	Corn oil
Canola oil	Donuts
Flax seed oil	French fries
Green leafy vegetables	Ice cream
Herring	Lamb chops
Lean meats	Margarine
Mackerel	Mayonnaise
Olive oil	Onion rings
Salmon	Potato chips
Sardines	Processed foods
Trout	Steak

GOOD FATS (HIGH IN OMEGA-3)	BAD FATS (HIGH IN OMEGA-6)
Tuna	Sunflower oil
Walnuts	Whipped cream
Whitefish	Whole milk

Extra-Credit Mental Aerobic Exercise

1. **A group of unexpected out-of-towners will be arriving at your home in 15 minutes, and they are absolutely starving. You have at your house everything from the left side column above (the good fats). As quickly as you can, create a dinner menu including as many food items from this list as possible.**
2. **Now, just as quickly, create a new menu using only items from the right column.**
3. **Afterward, imagine six assorted people in your life who you invite to a dinner party. Which of the two menus you just created would each of these people choose?**
4. **Do you notice the people who pick the omega-3 "good fat" menu are more likely to be concerned about their health, diet, and possibly even their brain aging?**

Catch of the Day

Scientists have shown that one of the omega-3 fatty acids, docosahexaenic acid, or DHA, which comes from fish oil, actually increases *acetylcholine*, the brain messenger critical to normal memory function but lost in Alzheimer's disease. People with deficient

DHA in their diets or low levels in the blood will experience learning difficulties and cognitive decline. These can and do improve when dietary DHA is high. Research indicates that omega-3 fatty acid capsules may improve memory difficulties and other symptoms in patients with Alzheimer's disease.

Recent studies suggest that omega-3-rich fish oil may benefit a person's mood as well as their memory, acting as an antidepressant and diminishing symptoms of hostility and aggression. Fish oil also has an antioxidant effect that fights against free radicals that can damage brain cells and decrease the brain's immune response (see later discussion), thus modulating the cell-damaging effects of inflammation. Because low-fat diets protect us against Alzheimer's disease, a healthy brain diet should include all kinds of fish, not just those high in omega-3 fats but fish that are considered low in their overall fat content, such as swordfish, snapper, sole, cod, catfish, flounder, perch, shellfish, haddock, and grouper.

Dr. David Heber and Susan Bowerman of the UCLA Center for Human Nutrition make a distinction between ocean-caught and farm-raised fish. This might sound like splitting hairs, but it's not. Farmed fish are fattier because they don't move around much, and their ratio of omega-3 to omega-6 is not as desirable because they don't eat the algae and other fish the way their free-swimming counterparts do in the ocean. Ocean-caught fish have less overall fat but more of the omega-3 fats because they are eating natural diets.

Healthy Brain Diet Tip

Eat fish at least twice a week.

It's Not the Sixties—Beware of Free Radicals

As we age, our brain cells undergo wear and tear from various oxidants known as *free radicals*. These free radicals are impossible to avoid—they are present in the air we breathe, the food we eat, and the water we drink. They perform useful functions in the body, but in surplus they can harm normal cells, wearing down their genetic material, or DNA. Brain cells, too, can suffer from this oxidative stress, a continual bombardment from chemical reactions in the environment and from within our own bodies. Through the DNA damage, this oxidative stress accelerates aging and promotes nearly all chronic age-related diseases from cancer to cataracts to Alzheimer's.

To keep oxidation in check, our bodies use antioxidants like vitamins C and E that combat the effect of free radicals. Recent studies show that people with low blood levels of these antioxidant vitamins have poorer memory abilities. Epidemiologists who've followed people over time in their communities while testing their memory and other cognitive performances report that those taking supplemental vitamin C and E tablets appear to have better memory abilities and less cognitive decline.

Dr. Martha Morris and her colleagues at Rush University and other centers looked at volunteers over age 65 years for four years. Although the usual percentage of people in that age group developed Alzheimer's disease as expected, not one of the subjects who regularly took the antioxidant vitamins C and E were among the group that developed the disease.

Unfortunately, these studies only recorded whether or not the participant was taking a supplement beyond just a daily multivitamin tablet, which generally contains about 30 units of vitamin E and 60 milligrams of vitamin C. The investigators did not determine the optimal vitamin supplement dosage for preventing Alzheimer's disease. Exactly how much your doctor might recommend for you is a matter of clinical judgment.

In a major study of Alzheimer's disease, the investigators chose a high enough dose of vitamin E—2,000 units daily—enough to feel assured that its antioxidant effects got to the brain. They found that this dose slowed down the advance of Alzheimer's disease by approximately seven months. Patients taking the vitamin were less likely to enter nursing homes or to develop severe symptoms for that period or longer. Many experts, therefore, recommend taking vitamin E at 1,000 units, twice daily, for severe memory loss as seen in Alzheimer's disease.

For people with only mild memory complaints who would like to benefit from vitamin E's antioxidant effects that might prevent future severe memory losses, deciding on an optimal daily dose is complicated by the vitamin's effect on the immune system. At low doses of 200 units each day, vitamin E may help reduce infections in older people, but at higher doses, it may have the opposite effect. Studies have shown that very high doses of vitamin E, for example, the equivalent of over 2,000 units daily, may suppress a person's immune response and limit the body's ability to fight off infections. Thus, doses above 1,500 units are rarely recommended except in cases of full-blown Alzheimer's disease.

In response to concerns about vitamin mega-doses, the National Academy of Science's Institute of Medicine recently recommended upper limits for antioxidant vitamins: 2,000 mg of vitamin C and 1,500 units of vitamin E in the natural d-α-tocopherol form (1,100 units in the synthetic dl-α-tocopherol form). The key is to find an effective and safe dose, while avoiding a mega-dose.

For healthy people who wish to take antioxidant supplements as part of their healthy brain diet, I recommend a daily dose of 400 units to 800 units of vitamin E, and 500 to 1,000 mg of vitamin C. Antioxidant foods and supplements not only help protect our brains but also protect our bodies against some forms of cancer, diabetes, and Parkinson's disease, as well as increase our immune defenses to colds and viruses.

Antioxidant Brain Food

Antioxidants occur naturally in many fruits and vegetables, and nutritionists have been touting their benefits for years. Dr. Jim Joseph of the U.S. Department of Agriculture found that laboratory animals fed on these natural antioxidant foods show better memory ability in finding their way through mazes and other tasks. Dr. Joseph encourages people to regularly eat antioxidant-rich foods such as strawberries, blueberries, raspberries, cranberries, broccoli, and spinach.

Researchers at Tufts University have devised a laboratory technique that measures the ability of different foods to counteract oxidative stress. Those foods with high "oxygen radical absorbency capacity," or ORAC, scores may protect our brain cells from the damage of oxidants—that of the free radicals, as mentioned earlier. Table 7.2 indicates some foods with potent antioxidant protection.

Table 7.2

THE TOP ANTIOXIDANT FRUITS AND VEGETABLES

FOOD	ANTIOXIDANT POWER*
	ORAC Units per 3½ Ounces
Prunes	5,770
Raisins	2,830
Blueberries	2,400

FOOD	ANTIOXIDANT POWER*
Blackberries	2,040
Cranberries	1,750
Strawberries	1,540
Spinach	1,260
Raspberries	1,230
Brussels sprouts	980
Plums	950
Broccoli florets	890
Beets	840
Avocados	780
Oranges	750
Red grapes	740
Red bell peppers	710
Cherries	670
Kiwis	600
Onions	450
Corn	400
Eggplant	390

**From U.S. Department of Agriculture Agricultural Research Service.*

The Tufts University experts recommend we all eat about 3,500 ORAC units each day—and just one cup of blueberries nearly accomplishes this goal. Most Americans and Europeans consume just over 1,000 ORAC units each day and generally don't get enough antioxidant foods in their diets. By simply doubling our average fruit and vegetable intake, we could each raise our diet's antioxidant power by 25 percent. Although during the last few decades Americans have successfully reduced their fat intake, their fruit and vegetable consumption remains relatively low.

The usual assumption that fresh is better than frozen does not necessarily hold true when it comes to the antioxidant capacity of foods. Studies of strawberries and blueberries show that the antioxidant properties of the frozen versions can actually be five times greater than the fresh varieties.

Tomatoes have been found to contain high concentrations of a particularly potent antioxidant called *lycopene.* Dr. David Snowdon of the University of Kentucky determined that women in their late seventies and eighties who had low blood lycopene levels showed decreased cognitive performance and a greater need for assistance in performing daily activities compared with women with higher lycopene levels. Eating foods rich in lycopene, such as tomato or V-8 juice, can dramatically increase the blood's antioxidant capacity. The UCLA human nutrition research group found that just six ounces of tomato juice increases lycopene blood levels by 40 percent. Mixed with some nice omega-3-rich olive oil, a little fresh basil, and some linguini, and we've got a brain-healthy lunch! Pass the salad, please.

Healthy Brain Diet Tip

Eat at least five servings of fruits and vegetables each day.

Dried fruits such as raisins and prunes are excellent sources of antioxidants; however, people concerned about calories might consider other sources because dried fruit tends to have a high caloric content. Tea, the second most consumed beverage worldwide, just behind water, is an excellent antioxidant source that does not contain calories. Tea is one of the few foods to contain significant amounts of the potent antioxidant known as *catechin*. Caffeinated green teas have very high catechin levels, as do caffeinated black teas brewed from bags.

A new approach that encourages people to eat foods with high antioxidant capacity emphasizes the color factor in fruits and vegetables. Dr. Jim Joseph in his book *The Color Code* and Dr. Dave Heber in *What Color Is Your Diet?* describe how *phytochemicals*—rich antioxidant dietary sources—are responsible for the colors in fruits and vegetables. Among these, anthocyanin makes a blueberry blue and has antioxidants that fight cancer; lycopene makes tomatoes red and protects our hearts and brains. Even the National Cancer Institute advises people to color their diets in its new Sample the Spectrum campaign.

Spice It Up

For many years I have preferred the dark, flavorful Dijon variety of mustard, while watching my kids squirt that bright yellow stuff on their hot dogs. Little did I know that the more colorful mustard version contained much higher concentrations of turmeric, a spice from the thick, rounded underground stems of a large-leaved herb cultivated in tropical countries. Turmeric is also the spice in curry powder and traditional Indian medicines used for thousands of years, and its active ingredient is *curcumin*.

Laboratory studies indicate that curcumin is a powerful antioxidant that inhibits those pesky free radicals and has anti-inflammatory actions besides. Scientists have found that curcumin relieves symptoms in arthritis sufferers and inhibits the growth of various cancer and tumor cells. Such encouraging curcumin effects led Dr. Greg

Cole and Dr. Sally Frautschy at UCLA and its affiliated Veterans Affairs Medical Center to study this ubiquitous spice's potential for preventing Alzheimer's disease. Their initial investigations of laboratory animals confirmed that curcumin not only suppresses oxidative damage to cells but also prevents loss of synapses—the connecting communicating terminals between brain cells—and decreases the deposition of amyloid protein and plaque burden in the brain.

Vitamins and Minerals for Keeping Brains Young

When a patient complains of a memory problem, one of the first things we do is to test their blood levels of vitamin B_{12}. Low levels of thiamine, vitamin B_{12}, or folic acid have been found to cause memory disorders. Some studies have shown that when Alzheimer's victims are treated with high doses of vitamin B_{12}, or folate, their memory abilities improve. Almost any vitamin deficiency will affect brain fitness and should be avoided.

Dr. Katherine Tucker, a nutritional epidemiologist from Tufts University, advises healthy older people to take a daily vitamin B_{12} supplement in addition to a daily multivitamin. Research has shown that 20 percent of people age 60 and older, and 40 percent of those over age 80, lose some of their ability to absorb vitamin B_{12}. Folate, or folic acid, is an antioxidant B vitamin that also offers a certain amount of protection against strokes, heart disease, and circulatory problems. In his long-term study of aging nuns, Dr. David Snowdon found that, at autopsy, the most extensive evidence of Alzheimer's disease was observed in the brains of those nuns who had the lowest concentrations of folic acid in their bloodstream.

Interestingly, folic acid supplements are recommended for women during their child-bearing years because a high enough blood level at the time of impregnation greatly diminishes the risk of having a baby born with spina bifida or other neural tube disorders.

Folic acid, as well as all antioxidants, is important to neural integrity and brain health throughout life, even before birth.

Recently, neuropsychologist Asenath La Rue of the University of New Mexico examined the mental and nutritional status of well-educated people ages 66 to 90, who were free of any major memory problems. Her study showed that the volunteers taking thiamin, riboflavin, niacin, and folate supplements scored better in abstract thinking tests. Also, volunteers with heightened blood levels of vitamin C scored higher in visual and spatial ability tests. These findings point to the potential brain-boosting effects of various vitamin supplements.

Further research has shown that older persons with low levels of thiamin, riboflavin, vitamin B_{12}, and vitamin C are more likely to experience anxiety, irritability, and depression. People who eat well-balanced meals generally don't develop deficiencies, and most doctors recommend daily multiple vitamins to ensure that such deficiencies do not develop.

It is also important for all of us to keep in mind the potential toxic effects of unnecessary vitamin mega-doses. This may be a particular problem with fat-soluble vitamins such as vitamins A, D, E, and K, which get stored in our body fat and can hang around in our bodies for weeks, months, or longer. When a little bit is good, a lot isn't always better.

Table 7.3 indicates the recommended daily allowance (RDA) of the vitamins we should all be taking—that is, their minimum doses according to the FDA (Food and Drug Administration); the safe dose or uppermost limit we can safely take each day; and some common food sources where we can get these vitamins and minerals without taking supplements.

Table 7.3

STAY HEALTHY WITH VITAMINS AND MINERALS			
NUTRIENT	RDA	DAILY SAFE DOSE*	FOOD SOURCES
Fat-Soluble Vitamins			
Vitamin A	5,000 IU	10,000 IU	Milk, eggs, leafy vegetables
Vitamin D	400 IU	800 IU	Milk, eggs, tuna, salmon
Vitamin E	30 IU	1,500 IU	Berries, vegetable oil, lettuce
Vitamin K	65–80 mcg		Leafy vegetables, fish oils, meat
Water-Soluble Vitamins			
Thiamin	1.5 mg	50 mg	Cereals, fish, lean meat, milk, chicken
Riboflavin	1.7 mg	200 mg	Cereals, milk, eggs, leafy vegetables, lean meat
Vitamin B_3 complex (Niacin)	18 mg	Nicotinic acid (500 mg); Nicotimamide (1,500 mg)	Cereals, lean meat, eggs
Pyridoxin (B_6)	2 mg	200 mg	Cereals, meat, bananas, vegetables
Vitamin B_{12}	3 mcg	3,000 mcg	Fish, lean meat, milk
Folate	400 mcg	1,000 mcg	Meat, leafy green vegetables

NUTRIENT	RDA	DAILY SAFE DOSE*	FOOD SOURCES
Water-Soluble Vitamins			
Vitamin C	60 mg	Over 1,000 mg	Citrus fruits, berries, vegetables
Minerals			
Calcium	800–1,000 mg	1,500 mg	Milk, cheese, green vegetables
Chromium	50 mcg	1,000 mcg	Whole grains, vegetable oils
Iron	10–18 mg	65 mg	Whole grain cereals, nuts, green vegetables
Magnesium	300–400 mg	700 mg	Whole grains, seafood, green vegetables
Selenium	50 mcg	200 mcg	Whole grains, seafood, eggs, meat
Zinc	15 mg	30 mg	Whole grains, sunflower seeds

From Hathcock (1997).

Another Supplement: Phosphatidylserine

Phosphatidylserine is a naturally occurring nutrient that exists in common foods such as fish, green leafy vegetables, soy products, and rice. This nutrient can be found in our cell membranes; in fact, approximately 10 percent of the fatty component of our brain cell membranes consists of phosphatidylserine.

Scientists have found that phosphatidylserine can increase neurotransmitters that improve memory and concentration, and animal studies indicate that it slows age-related memory decline. These encouraging observations have led to studies testing the effectiveness of phosphatidylserine as a supplement to augment recall abilities in people with mild age-related memory complaints.

Dr. Tom Crook, a neuropsychologist formerly with the National Institute of Mental Health, along with other investigators, has shown that people with age-associated memory impairment score better on memory and learning tests after taking phosphatidylserine when compared with those taking a placebo.

This nutrient may indeed be effective, and perhaps sixty or more studies have demonstrated this modest but positive benefit. A limitation of these studies, however, is their relatively brief duration, ranging from six to twelve weeks, raising the possibility that the benefit may be not be long-term. It is certainly possible that phosphatidylserine has a long-term beneficial effect, but it has never been systematically studied beyond twelve weeks.

Doctors who recommend phosphatidylserine suggest that people begin with 100 to 150 mg twice a day and after several months they drop the dose to only 50 mg twice a day for maintenance. No side effects have been reported.

The Ups and Downs of Caffeine

Every morning, millions of us drag ourselves out of bed, blurry-eyed, empty mug in hand, and grope our way to the coffeepot before we can imagine beginning our day. Coffee consumption exceeds 100 billion cups per year in the United States alone, where 80 percent of the adult population drinks coffee or tea daily—making caffeine our most commonly used drug. We also get caffeine in our diet from sources that some people are not aware of, including chocolate and some sodas.

Too much caffeine increases cholesterol levels, may increase the risk of heart attacks, and is associated with urinary bladder cancer and high blood pressure. Caffeine also increases risk for bone thinning from osteoporosis. Acute caffeine intoxication causes rapid heart rate and can pose health hazards for cardiac patients.

Caffeine has both positive and negative effects on brain fitness. On the positive side, it diminishes fatigue, increases alertness and attention, and improves mood. We all know of the pick-me-up we get from our morning java. Systematic studies show that in the short term, caffeine can improve learning and recall abilities. In the Honolulu Heart Program, a thirty-year study of 8,000 people, the risk of developing Parkinson's disease was five times lower in coffee drinkers compared with those who did not drink coffee.

On the negative side, extended caffeine use can cause irritability, insomnia, and anxiety. Because caffeine's effects are short-acting, suddenly interrupting the caffeine habit can cause withdrawal symptoms. Caffeine withdrawal usually begins twelve to twenty-four hours after the last exposure, with symptoms peaking in the first forty-eight hours but sometimes lasting up to two weeks. Headache, fatigue, poor concentration, and depression are common complaints when we can't get our daily caffeine fix. Just ask my wife—but not before she's had her morning coffee. The single greatest cause of post-operative headache is caffeine withdrawal. In order to prevent post-operative headaches, some surgeons have been known to actually add caffeine to the intravenous fluids of patients who cannot drink liquids after surgery.

For those of us who do consume caffeine, we need to be aware of the side effects and avoid overuse and withdrawal symptoms. Table 7.4 lists the caffeine content of some common foods and drugs to help you to keep track of how much you're taking each day.

Table 7.4

CAFFEINE CONTENT OF SOME COMMON DRINKS, FOODS, AND DRUGS	
Brewed coffee (6 oz)	100 mg
Decaf coffee (6 oz)	4 mg
Instant coffee (6 oz)	70 mg
Tea (6 oz)	40 mg
Caffeinated soft drink	45 mg
Chocolate candy bar (40 gm)	10 mg
Anacin	32 mg
Excedrin	65 mg
No-Doz	100 mg

Sources: National Coffee Association, National Soft Drink Association, Tea Council of the USA, and Barone and Roberts (1996).

Sugar on the Brain: How Sweet It Is

Sugar, or glucose, is the brain's main energy source, and blood sugar levels affect both mood and memory. Unlike other cells in our bodies, brain cells cannot convert fats or proteins into glucose, so they depend on daily dietary sugar for optimal functioning and survival.

When our blood sugar levels drop too low, many of us tend to feel lethargic, irritable, and may have difficulty learning new information. But give us a meal, power bar, or glass of juice, and our moods perk right up, as do our memory and concentration abilities.

Our brains don't function well when blood sugar is too high, either. Studies of both animals and humans have consistently shown that abnormally low or high blood sugar levels will affect memory and learning abilities. The reason may involve the brain messenger or neurotransmitter acetylcholine, a neuron communication link important for normal memory performance.

The acute effects of sugar on the brain are well documented. After drinking carbohydrate-spiked lemonade, volunteers show better memory performance and mental flexibility than after drinking saccharine-sweetened lemonade. Similar results have been documented in patients with Alzheimer's disease.

When I was growing up, my father always encouraged my sisters and me to eat breakfast, "the most important meal of the day!" He never backed up this statement with facts, but I later learned that his was indeed good advice. Breakfast—breaking the fast of our nighttime sleep—increases blood sugar levels and leads to greater mental clarity during the day. Studies of elementary school students show improved academic performance and behavior when they eat breakfast, and adults who eat breakfast maintain higher blood sugar levels, quicker recall, and better overall memory performance than those who skip it.

Diabetes: When Your Blood Gets Too Sweet

Although sufficient blood sugar keeps our brains working optimally and a quick glucose fix can give us an immediate memory and concentration boost, many of us suffer from chronically high blood sugar levels. If sustained over months and years, these high blood glucose levels can lead to a pre-diabetic state and possibly impair memory and other mental abilities.

After a meal, our blood sugar increases, which triggers the pancreas to produce insulin, the hormone that facilitates sugar, or glucose, getting into cells where it is needed for energy. If, however, we are constantly experiencing repeated, sharp spikes of blood sugar, the pancreas can become overworked and eventually produce less effective insulin. This can cause the body to become insulin-resistant or unable to use insulin effectively, which puts one at risk for non-insulin-dependent diabetes, or type 2 diabetes, as well as high blood pressure and circulatory problems affecting the brain. Arteries can become stiffer, restricting blood flow to the brain.

About 16 million people in the U.S. suffer from diabetes, an increase of nearly 40 percent during the last decade. Diabetes can quadruple the rate of heart disease and stroke. The chronic high blood sugar levels of diabetes have also been linked to lower intellectual performance. Diabetics have an increased risk for developing severe memory loss associated with aging, including Alzheimer's disease and other types of dementia. Genetic predisposition can partly determine our susceptibility to high blood sugar and diabetes, but a person's daily dietary habits play a major role.

The good news is, even minor changes in our diets and other lifestyle areas can have a strong impact on our risk for diabetes. Dr. Jaakko Tuomilehto and his colleagues at the National Public Health Institute in Finland found that losing as few as ten pounds, eating a healthy diet, and exercising regularly can reduce the risk for developing type 2 diabetes by more than 50 percent.

Because our brains need a steady flow of sugar to keep them optimally fit, maintaining an even glucose level in the brain and avoiding blood sugar fluctuations should be everyone's goal. We can begin to protect our brains from the onslaught of chronic sugar overload and insulin surges by attempting to avoid foods that spike blood sugar and, in turn, cause the pancreas to pump out more insulin.

The sugars we eat are technically carbohydrates and they come in two forms: simple sugars, such as sucrose or table sugar, and complex sugars or starches, including fruits, vegetables, milk, and cereals. When asked why the rate of diabetes has risen sharply over the last few decades, experts point to the changes in our diets. In contrast to our ancestors' natural sources of carbohydrates—fruits and vegetables—

many people today eat foods containing refined sugars and processed flour. These newer, less nutritional forms of carbohydrates can cause rapid rises and subsequent falls in blood sugar levels, which our bodies were not designed for.

In recent years, scientists have begun studying the actual blood sugar responses to a variety of foods. They are then able to compare the food's true physiological effect on blood sugar levels, the *glycemic index*. This index ranks foods from 0 to 100, indicating whether the food raises blood sugar levels dramatically, moderately, or minimally. This research debunks many old myths about carbohydrates. First, starchy foods like bread, potatoes, and some types of rice are digested and absorbed quickly, rather than slowly. Second, foods with lots of sugar, like candy and ice cream, do not dramatically increase blood sugar but lead to low or moderate blood sugar responses, even lower than bread. Carbohydrates with a high glycemic index tend to decrease the good form of HDL cholesterol and to increase the risk for diabetes, insulin resistance, and heart disease. They also increase hunger and promote overeating and obesity.

While avoiding high-glycemic-index foods, we should also try to eat the low-glycemic-index foods that don't cause peaks and valleys in blood sugar levels but instead lead to gradual rises and falls in blood sugar levels. Low-glycemic-index "carbs" increase the good form of HDL cholesterol, tend to curb appetite, and help us to burn off fat. Exercise physiologists have found that eating low-glycemic-index carbs before sustained, strenuous exercise improves physical performance.

A recent study of nearly 36,000 women initially free of diabetes found that those eating low-glycemic-index grains, particularly whole grains and cereal fiber, had a lower risk for developing diabetes. Nutritional experts recommend the consumption of three servings of whole grains each day. On average, most Americans consume less than one serving.

In her book *The Glucose Revolution*, Dr. Jennie Brand-Miller of the University of Sydney and her colleagues provide a glycemic index table that ranks common foods according to how much they spike blood sugar levels. This index is an average from several studies of the foods' physiological effects. The glycemic index is independent of serving size, so you can eat more of a low-glycemic-index food and experience the same blood sugar levels you would from eating less of one with a high glycemic index. Table 7.5 lists some common foods ranked according to glycemic index.

Table 7.5

HOW MUCH DO SOME COMMON FOODS SPIKE BLOOD SUGAR?	
Minimal (Glycemic Index <40)	
Apple	Lima beans
Apricots, dried	Nonfat yogurt
Cherries	Peanut M&M's
Fettuccine	Peanuts
Kidney beans	Skim milk
Lentils	Soybeans
Low (Glycemic Index 40–54)	
Baked beans	Orange
Bran cereal	Orange juice
Canned chickpeas	Oatmeal
Cooked carrots	Potato chips

Low (Glycemic Index 40–54)	
Chocolate bar	Spaghetti
Grapes	Unsweetened apple juice
Moderate (Glycemic Index 55–70)	
Angel food cake	Natural muesli cereal
Bananas	Oat bran cereal
Brown rice	Pineapple
Canned corn or beets	Sourdough bread
Croissant	Potatoes
Honey	Whole wheat bread
Ice cream	White bread
High (Glycemic Index 71–84)	
Bagels	Bran flakes
Jelly beans	Pretzels
Cocoa Puffs	Puffed wheat cereal
Cheerios	Corn flakes
Raisin bran cereal	Total cereal
French fries	Vanilla wafers
Maximal (Glycemic Index >85)	
Dried dates	Instant mashed potatoes
French baguettes	Instant rice

The groupings in the table are meant as a guide to carbohydrate choices. We generally don't know the specific effect on blood sugar when we sit down to eat a particular meal because we usually eat foods in combinations that tend to minimize blood sugar spikes from high-glycemic-index foods. Highly acidic foods like vinegar also will minimize blood sugar spikes. Dr. Brand-Miller found that lemon juice and vinegar, particularly red wine vinegar, have this effect, which she attributes to an acidic food's tendency to slow the digestive process (see Tips box). Such acidity may also explain why sourdough bread has a lower glycemic index than some other breads. Also, the lactic acid found in yogurt may explain its tendency to minimize blood sugar spikes.

Tips for Avoiding Blood Sugar Spikes

- **Eat fresh fruits and vegetables.**
- **Avoid instant rice.**
- **When you eat foods that tend to spike glucose, combine them with foods that don't.**
- **Avoid processed foods.**
- **Add vinegar or lemon juice when you can.**
- **Eat small meals and avoid infrequent large meals.**
- **Eat a healthy breakfast every day.**

Stress Eating

What we eat and how we eat it has tremendous emotional meaning in our lives. Eating is often a symbol of love—mothers express anxiety or feelings of rejection when their children turn away their food, and the tradition of "breaking bread" is an important social and business ritual. Anxiety and stress can have a profound impact on some people's ability to eat sensibly and make it difficult to maintain a healthy brain diet.

Many people lose their appetite under extreme stress, while the opposite problem is a complaint for many of today's weight-conscious baby boomers. Stress often triggers impulse eating or perhaps leads you toward old habits, like downing your first grader's leftover potato chips while driving home from the office.

Have you ever picked up the phone, suffered through an unpleasant phone call, hung up, and noticed that eight Oreo cookies have disappeared? Do you find yourself guiltily hiding the evidence of eating your child's candy bar or ice cream sandwich when your day has overwhelmed you? Nearly everyone has experienced some form of stress eating at some time. The phenomenon generally has two components: (1) a stressful event to trigger binge eating, and (2) conveniently available foods, often processed foods or desserts.

Lisa E., a 45-year-old accountant, became adamant about starting a memory-training and mental stimulation program due to the recent changes in her recall abilities. Her symptoms had gotten markedly worse during the last tax season, and she didn't like it when both her boss and her 16-year-old son started making jokes about her middle-aged memory lapses.

She came to our clinic having already picked up some memory-training skills on her own, but wanting to learn more and hone her techniques. The results of Lisa's initial evaluation were normal, except for a borderline elevation in blood sugar, so we referred her to a nutritionist, who provided dietary guidelines to help her control her blood glucose levels. She agreed to start a low-fat diet, high in omega-3 fatty acids and antioxidant foods, and agreed to avoid high-glycemic-index foods. She also began a memory-training course and a mental aerobics program.

After eight weeks, Lisa noted some memory improvement, but her blood test still showed slightly elevated sugar levels. She swore she had been true to the diet, shopping at the health food market, buying only the foods on the list, and preparing them correctly.

What Lisa failed to mention were the donuts, sweet rolls, and frosted flakes she had at home for her son, as well as the constant bombardment of bagels, breakfast bars, and almost daily birthday cakes surrounding her at work.

When Lisa's nutritionist finally convinced her to keep a daily food log, diligently recording everything she ate for a week, Lisa learned the tragic truth she had been hiding, even from herself—she was a closet sugar addict. In moments of elevated stress, her unconscious sugar cravings would emerge. Like a vampire at nightfall, Lisa would reach out and devour the closest sugary snack food available—usually without even realizing it.

Thanks to her food log, Lisa learned to grapple with her sugar problem and get her blood glucose levels under control. She was able to keep her memory-training work on track and her job performance improved. Her boss gave her a raise, which was no joking matter, and the only person to complain was Lisa's son, who had to give up the donuts and frosted flakes for the healthier, natural alternatives Lisa now bought and enjoyed.

Carbohydrates with a high glycemic index are the usual culprits when it comes to stress eating. There's something extremely satisfying in that fleeting, momentary, crunchy, munchy sugary experience. However, the resulting insulin spikes followed by subsequent blood sugar crashes often leave one feeling famished and can lead to additional overeating as well as other serious problems.

Although we can't completely eliminate stress from our lives, we can follow some practical tips to avoid unhealthy stress eating (see box).

Practical Tips to Avoid Stress Eating

- **Eliminate your favorite unhealthy stress foods from your house, car, and office.**
- **Keep baggies of fresh-cut vegetables in convenient places where stress eating most often occurs—near the kitchen telephone, at the office desk, in the car, etc.**
- **Avoid processed-food snacks. Instead, if you need a quick snack, substitute "brain snacks": power bars, sourdough croutons, blueberries, and strawberries.**
- **Keep bottled water nearby at all times—when stress hits, take a swig.**
- **Set reminders—post a sign in each of your stress-eating spots (near the kitchen telephone, workplace computer station) reminding you to "Relax and Eat Healthy."**
- **When you catch yourself in a stress-eating mode, put yourself on pause: take a deep breath, toss out that cookie or donut, take a break from the stressful situation, and stretch.**
- **Develop other skills to reduce stress by reading Chapter 4.**

Start Dishing Up the Smart Food

Eating a healthy brain diet is no more complicated than any physically healthy diet. Getting enough antioxidants, the right fats and carbs, and limiting calories are all easier than it may seem at first read. Probably the hardest part is just getting started. Once you've passed that hurdle, the rest is cake (only an expression, of course).

Dr. David Heber of UCLA emphasizes eating not just healthy foods, but tasty foods as well. He recommends fruits and vegetables; high-fiber breads, cereals, and grains; and low-fat animal proteins (e.g., skinless chicken, fish, skim milk products). Dr. Heber encourages people to use herbs, spices, garlic, chili peppers, avocados, nuts, seeds, and olives as taste enhancers. Emphasis on taste as well as health usually helps keep us on our healthy brain diets longer, hopefully for the rest of our lives.

When it comes to food and brain health, setting reasonable goals and being patient can be your greatest assets. By following some basic guidelines, your brain fitness will likely improve quickly, and for the long run.

The Basic Elements of a Healthy Brain Diet

- **Eat a low-fat diet.**
- **Stay aware of overall caloric intake.**
- **Avoid stress eating and late-night snacks.**
- **Toss away your yo-yo diet plan.**
- **Avoid processed foods and high-glycemic-index carbs.**
- **Eat foods rich in omega-3 fats.**
- **Avoid omega-6 fats.**
- **Remember you have choices: sourdough bread rather than French rolls, olive oil rather than corn oil.**
- **For a daily antioxidant boost, eat fruits and vegetables, and drink tea. Try frozen or fresh blueberries for a snack food.**

- **Avoid too much caffeine.**
- **Drink water throughout the day; shoot for six or more glasses daily.**
- **Take multivitamins as well as vitamins E and C and folate supplements.**
- **Create meals that are healthy and tasty—try using herbs, garlic, and spices to enhance taste.**

The following is a sample of several days from one person's Weekly Healthy Brain Diet Worksheet for you to examine. This individual starts out strong during the workweek, but has more difficulty as the temptations of the weekend rear their ugly head. The process of keeping a log and viewing overall progress throughout the week can help us to focus on times and situations when we need to work harder to stay on a healthy diet. This chart may serve as a guide to help you create a similar chart for yourself.

Healthy Brain Diet Worksheet

WEDNESDAY	THURSDAY	FRIDAY	SATURDAY
Breakfast 1 cup of coffee Vitamins Egg white Bran muffin, orange	**Breakfast** 1 cup of coffee Vitamins Egg white Sourdough toast	**Breakfast** 1 cup of coffee Vitamins Egg white Sourdough toast	**Breakfast** 1 cup of coffee Vitamins Onion bagel, lox, cream cheese
Morning Snack Low-fat muffin Carrots/celery 1 cup of coffee	**Morning Snack** Power bar 1 cup of coffee	**Morning Snack** Power bar 1 cup of coffee	**Morning Snack** Bagel & light cream cheese 1 cup of coffee
Lunch Diet soda Turkey & swiss on rye Raw veggies Yogurt	**Lunch** Green tea Chicken soup Peach Salad lo-cal dressing	**Lunch** Diet soda Turkey & swiss Oatmeal cookie	**Lunch** Diet soda Turkey dog Sauerkraut
Afternoon Snack Blueberries String cheese	**Afternoon Snack** Banana Yogurt	**Afternoon Snack** Blueberries String cheese	**Afternoon Snack** Power bar String cheese
Dinner 1 glass red wine Salad, olive oil & wine vinegar Pasta/turkey meatballs Steamed veggies Toast	**Dinner** 1 glass red wine White fish Steamed veggies Salad, olive oil & wine vinegar	**Dinner** Tomato soup, croutons Salmon, rice Steamed veggies Sourdough toast	**Dinner** 2 glasses red wine Steak & fries Tomato soup Dessert wine
Night Snack Frozen fruit bar Raisins, apple	**Night Snack** Frozen fruit bar Orange	**Night Snack** Ice cream Popcorn	**Night Snack** Leftover cake
Total water count 7 glasses	**Total water count** 6 glasses	**Total water count** 6 glasses	**Total water count** 5 glasses

Chapter Eight

Choose a Lifestyle That Protects Your Brain

If I'd known how old I was going to be I'd have taken better care of myself.
—Adolph Zukor, founder of Paramount Pictures, before his 100th birthday

Our western approach to medicine has traditionally emphasized curing illness rather than maintaining wellness. Nearly all long-term studies on aging and memory have focused on markers that predict decline and loss. This trend, however, has been slowly changing. Scientists have begun focusing their investigations on successful aging and late-life health.

Successful aging means not only living longer but living *better*—avoiding disease, remaining engaged in activities, and maintaining optimal physical and mental health. During the past decade, the MacArthur Foundation supported a study that took an innovative approach to aging and stressed positive rather than negative outcomes.

Dr. Robert Kahn and Dr. John Rowe summarized the MacArthur findings in their book *Successful Aging* and gave baby boomers a reason for optimism: the lifestyle choices we make early in life determine our health and vitality as we age, even more than heredity and genetics. Only about a third of what determines successful aging is already programmed through our individual genetic codes. The other two-thirds result from our environments, and in large part, the choices we make that become our lifestyles.

Many of the lifestyle changes we need to consider for keeping our brains young are the same habits that will help us to maintain physical health and fitness. The U.S. Surgeon General, Dr. David Satcher, advises Americans to follow his prescription for healthy living with recommendations that include moderate physical activity—at least 30 minutes a day, five days a week; eating at least five servings of fruits and vegetables each day; and avoiding tobacco, illicit drugs, and alcohol abuse. If there were a comparable office known as Surgeon General for Brain Fitness, I suspect it would issue similar recommendations.

Dr. Karen Ritchie, an epidemiologist from Montpellier, France, studied the mental status of a woman at age 118. Her memory ability was comparable to that of a normal 80-year-old. Although she had a genetic predisposition to longevity, her lifestyle likely kept her brain young: she was educated, remained mentally and physically active, and ate the typical diet of Provence, France—olive oil, fresh vegetables, and fish.

I am often asked at what age it becomes too late to change bad habits, start taking care of one's body, and thereby help to protect one's brain. Allow me to say it clear and loud: it is never too late. As soon as you start to change your lifestyle for the better, you'll begin to repair yesterday's damage. A previously sedentary 40-year-old who begins a walking program of just 30 minutes a day, four days a week, can achieve the same risk of heart attack after six months of conditioning as a 40-year-old who has exercised conscientiously for decades.

Physical Exercise and Brain Fitness

Recent discoveries show that physical activity and aerobic conditioning promote brain fitness. Armed with convincing evidence from large-scale, long-term human studies, as well as experiments in laboratory animals, scientists are now recognizing that physical activity apparently protects the memory centers of the brain.

Most of us who do aerobic exercise on a regular basis generally do so to maintain physical stamina, health, and fitness. Physical exercise can also enhance our mental state by increasing the circulation of *endorphins*—hormones released in our brains after exercise that have immediate benefits on mood and memory—the body's own internal antidepressant. Regularly scheduled aerobic exercise, with its accompanying mildly euphoric endorphin "boost," also helps maintain maximum long-term brain health.

A well-balanced exercise program usually includes some toning and stretching, which allows people to avoid injury while they build stamina. The actual aerobic part of our exercise routine gets our hearts pumping faster, our lungs breathing deeper, and if we continue these activities on a regular basis, they can help reduce our risks for age-related illnesses like heart attacks and strokes. Many experts recommend walking as one of the safest and most effective forms of aerobic exercise. The MacArthur Study of Successful Aging noted that older adults who walked 45 minutes, three to four times a week, doubled their endurance level after a year.

A convincing case can be made for the mental benefits of aerobic exercise, although a Heisman Trophy is not a 100 percent guarantee against developing Alzheimer's disease. The disease has been known to strike people who have achieved remarkable feats of physical fitness, including triathletes, long-distance runners, and championship tennis players.

Recent studies, however, indicate a definite link between physical activity and staving off Alzheimer's disease. Small laboratory animals exposed to exercise—running on wheels and treadmills—show formation of new blood vessels and nerve cell communication sites, or synapses, in the brain. Dr. Fred Gage and his co-workers at the

Salk Institute in La Jolla, California, have found that adult mice exercising regularly on a running wheel developed twice as many new brain cells in the hippocampus compared with mice in standard cages. The scientists speculate that running might increase the flow of oxygen and nutrients to brain tissues or release special growth factors that promote nerve cell growth. These landmark studies also contradict the old myth that new cell growth does not occur in adult brains. Physical exercise appears not only to keep brain cells alive but also to grow new neurons.

Dr. Robert Friedland and his associates at Case Western Reserve University studied over 500 people to determine their physical activity levels. The volunteers who had been physically active between the ages of 20 and 60 were three times less likely to suffer from Alzheimer's disease later in life, and these activities ranged from gardening a few times a week to racquetball to daily jogging.

Researchers have found that physical exercise provides benefits in mental performance, regardless of age. In fact, provided people do not *over*-exert themselves, these benefits can be observed immediately following exercise. Long-term cognitive benefits have also been noted with continued physical conditioning.

Recent investigations suggest that the greatest short-term cognitive benefits of aerobic fitness involve task solving, or what psychologists call executive control: making plans, scheduling and carrying out activities, coordinating events, and controlling emotional outbursts or "keeping a poker face." These processes generally involve the frontal or prefrontal region of the brain, often considered the more highly evolved brain region. The frontal cortex of animals such as cats, turtles, and squirrels is clearly less developed than that of humans, and humans have the greatest capacity in this area of the brain of any animal.

As we age, this frontal lobe gradually shrinks in size. In addition, brain activity levels in this frontal area gradually decrease at a faster rate than the rest of the brain. Middle brain regions, bridging the front and back of the brain, remain at a constant level of activity throughout a normal human lifespan. These areas involve basic functions such as sensation and motor control, and remain normally active, even in patients with advanced Alzheimer's. Many experts agree that this frontal cortex of the brain is the area most likely to benefit from physical aerobic training.

Tennis players, runners, and other athletes in their sixties and older have faster mental responses and reaction times than those of non-exercisers of the same age. They also outperform their inactive counterparts on tests of reasoning, memory, attention, and intelligence. Studies of aging athletes, however, may reflect some other advantages associated with being physically active, such as a healthy diet, good genetic predisposition, or use of anti-inflammatory drugs. Research data vary according to the age of the volunteer. In fact, the older the study subject, the more prominent the mental benefits of physical exercise. Also, if a person is physically fit at the outset of testing, it may be more difficult to measure the mental benefits of working out.

Dr. Arthur Kramer and his associates at the University of Illinois demonstrated the mental benefits of physical aerobic exercise in a six-month study of healthy adults between ages 60 and 75. Divided into two groups, one followed an aerobic walking program, while the comparison anaerobic group followed a toning and stretching program. The aerobic exercise group learned basic principles and guidelines for exercise programming, including an adequate warm-up and cool-down period, and increases in exercise duration and energy expenditure in gradual and progressive increments. Subjects also received instruction on how to avoid exercise-related injury. The investigators predicted that this program would improve brain function in the frontal lobe.

The aerobic group worked out three times a week beginning

with 10- to 15-minute sessions, increasing by a minute each session, and eventually building up to 40 minutes per session. The control group worked on a similar schedule, but instead of aerobics, they were instructed in techniques for stretching their range of motion. Stretches were held to the point of slight discomfort and involved all large muscle groups throughout the body.

As the scientists predicted, mental tasks involved in executive control—monitoring, scheduling, planning, inhibition, and memory—improved in the aerobic group but not in the control group. The benefits on mental attention were particularly striking.

Aerobics are not the only aspect of exercise that keeps our brains young, however. Geriatricians who studied weight training in older adults found that after just three months of training, older men could double the strength of their quads—the front thigh muscles—and triple the strength of their hamstrings—the back thigh muscles. Weight training also enlarged their muscles. The older adults not only increased their strength, they dramatically improved their balance in just a few months.

Increased muscle tissue allows the body's metabolism to function at a higher rate throughout the day, which in turn, uses up more calories. The resultant weight control can help prevent physical illnesses related to obesity, including hypertension, stroke, and diabetes, all of which can accelerate brain aging.

Despite the wide variety of sports and fitness programs available today, many baby boomers avoid them entirely, allegedly too busy working, carpooling, and caring for parents and children simultaneously. Although it is challenging to squeeze a physical activity routine into a crowded week, it can be done, and it should be a priority.

Once you develop the exercise habit, you may quickly get hooked on the endorphin blast, and then the other benefits literally *show up*—in the mirror and in how good you feel. For those who absolutely don't have the time, start by popping a bit of exercise into your normal daily routine: take the stairs instead of the elevator; choose a 5- or 10-minute brisk walk over another coffee break, and involve your mate or a friend in an outdoor weekend activity instead of lounging all Sunday in pajamas watching a *Twilight Zone* marathon.

Any physical exercise program should include a series of stretching and toning exercises, along with a good aerobic component. The key is to blend exercise into your lifestyle and make it a part of your daily routine. Even if you can only spare 10 or 15 minutes a day at first, make the best use of those 10 or 15 minutes and do it every day if possible.

Simply following some general tips for beginning a physical exercise program can help keep your brain young (see box).

Tips for Starting an Exercise Program for Healthy Brains

- **Be sure to include adequate warm-up and cool-down periods.**
- **Always include stretching and toning of all large muscle groups to increase flexibility and avoid injury.**
- **Walk with friends on a regular basis for both the social and physical benefits.**
- **Learn about exercise-related injuries and how to avoid them. Wear proper footwear and clothing to avoid injury and temperature extremes.**
- **Increase your exercise duration and energy expenditure gradually and progressively over time. You don't have to become a triathlete to maintain brain fit-**

ness—moderate but regular conditioning is more than enough.

- **Incorporate both an aerobic and a weight-training component to your program.**
- **Slip some type of physical activity into your daily routine: take the stairs instead of the elevator, a brisk walk instead of a donut break.**
- **Choose sports and activities that you enjoy because you are more likely to continue them in the long run. If walking around the neighborhood bores you, try a treadmill or stationary bike so you can read or watch TV while you work out.**
- **Check with your physician when getting started, especially if you have a physical illness that an exercise program could affect.**

Watch Your Head

When choosing an aerobic fitness program, I advise my patients to avoid those that increase risk for head trauma. More than 5 million Americans have suffered from some type of traumatic brain injury, and nearly all those injuries could have been prevented. The scientific evidence points to the obvious: avoid head trauma—both mild and severe—to protect your brain from cognitive decline. Wearing seat belts, choosing a designated driver who is not drinking, and wearing helmets when riding bikes or doing sports are critical to protecting our brains.

Dr. Richard Mayeux and research associates at Columbia University found that people who have blacked out for an hour or more following a head trauma have a twofold increased risk for developing Alzheimer's disease down the road. If such a person also has the

APOE-4 genetic risk for Alzheimer's disease, their overall risk for the disease increases to tenfold.

Dr. Brenda Plassman and her co-workers at Duke University studied medical records of veterans who had suffered varying degrees of head trauma, dating back as far as fifty years. They found that veterans with only moderate head injury—loss of consciousness or post-traumatic amnesia for more than 30 minutes and less than twenty-four hours—had a twofold increased risk for Alzheimer's disease over those without a history of head trauma. And, the Alzheimer's disease risk increased in correspondence with the seriousness of the injury—those who had been hospitalized or suffered amnesia for more than twenty-four hours had a *fourfold* increased risk of developing Alzheimer's disease sometime in the future.

Nearly all these studies looked at moderate to severe head trauma, but many experts are convinced that even milder forms of repeated head injury could accelerate brain aging. Recent studies have focused on the memory effects of mild but repetitive brain injuries caused by contact sports. Dr. Erik Matser and his team at St. Anna Hospital in the Netherlands compared amateur soccer players in their mid-twenties to same-aged swimmers and runners who were less likely to suffer head injuries. Over 30 percent of the soccer players suffered from memory impairments, while less than 10 percent of the swimmers and runners had similar impairments. Although the memory impairments in these athletes were mild, it does raise concern over possible risk for future progressive decline.

At UCLA, Dr. David Hovda and Dr. Marvin Bergsneider performed PET scans on patients who had recently experienced relatively mild concussions. They found the brain activity of patients with only mild concussions was similar to that of comatose, severely brain-injured patients. Dr. Hovda noted that although a person may be able to walk, talk, and appear normal and alert after a concussion, their brain may not be functioning normally.

With this mounting evidence in mind, if you're the quarterback on your company's team and getting sacked two or three times every Sunday, don't be surprised if you forget your first two meetings on Monday morning. You may want to consider tennis.

Dr. Paul Satz at UCLA describes a *brain reserve capacity* that varies among individuals, giving each person a different threshold of brain injury required before memory loss and other problems emerge. A high degree of reserve capacity will protect the brain, so that a mild blow to the head might cause no symptoms in one person while causing severe injury and cell loss in another—depending on the individual's brain reserve capacity threshold. A redundancy in the neuronal networks may explain such reserve. Satz's theory is consistent with the idea that multiple small injuries have a cumulative effect in whittling away brain reserves until a certain level or threshold of cumulative damage is reached and symptoms become apparent.

Other evidence supports this notion of brain reserve. Dr. James Mortimer and associates at the University of South Florida looked at head circumference as a reflection of neuronal numbers and density of their interconnections. His group, as well as others, found that the size of a person's head does indeed predict their future risk for getting Alzheimer's disease—big heads have lower risks. Studies using brain scans to measure brain size further supported this concept. Dr. Peter Schofield and co-workers at Columbia University found that the onset of Alzheimer's disease is delayed by four months for every one square centimeter increase in brain size.

After head trauma, the brain immediately responds by forming amyloid plaques—those same collections of cell decay that indicate a diagnosis of Alzheimer's disease—yet another link between head injury and the disease. Remarkably, although the plaques that develop following an injury are usually more obvious in older than in younger patients, such plaques have been observed in patients as young as 10 years of age.

The APOE-4 genetic risk for Alzheimer's disease contributes to the cognitive decline following head injury. UCLA investigator Dr. Barry Jordan found that possession of this Alzheimer's risk gene was associated with more severe neurological deficits in boxers. Also, head-injured patients with a high dose of the risk gene have greater amounts of amyloid plaque deposition in their brains. The science tells us that people with the APOE-4 gene have an even greater reason to avoid sports and occupations that involve a high risk for head injury, such as boxing, football, soccer, race car driving, movie stunt work, and crash-helmet testing.

Just Say No to Smoking

Everybody knows that smoking is bad for us—it can lead to lung and other cancers, heart disease, stroke, and numerous other disorders. But many people are unaware of the damage smoking does to the health of our brains. Studies show that smokers have a definite increased risk for Alzheimer's disease. Dr. Richard Mayeux's group at Columbia University studied a large number of older adults and found that smokers had a twofold greater risk of getting Alzheimer's disease than those who never smoked. However, when people quit smoking, at whatever age, they were able to slightly reduce their risk.

The U.S. Surgeon General, Dr. David Satcher, advises all smokers to quit. Once a person quits smoking, the benefits emerge rapidly. The body's carbon monoxide levels drop dramatically, and within a week, the risk of dying from a heart attack begins to decline. Five years later, that person's heart attack risk is similar to that of someone who never smoked. Treatment programs that include counseling, as well as educational and emotional support, do succeed when participants make a reasonable effort. Some intensive treatment programs have success rates for long-term abstinence approaching 50 percent. Nicotine patches and gum have proven effective as well, particularly when used along with other treatment approaches. Antidepressants like bupropion (Wellbutrin) can also help some people to quit smoking. Of course, nothing will work if the smoker does not truly desire to stop, for whatever reason. Knowing the connection between smoking and cognitive function is just one more good reason to quit.

Researchers have been testing nicotine patches because of the potential benefits of this neurotransmitter in diseases raging from schizophrenia to Tourette's syndrome. Because nicotine receptors decline in the brains of Alzheimer's victims, use of nicotine-enhancing drugs is one treatment strategy for age-related memory loss (Chapter 9). Dr. Paul Newhouse of the University of Vermont has tested a synthetic form of nicotine on a small number of Alzheimer's patients and found an improvement in their learning and memory abilities.

Although nicotine may benefit some brain receptors involved in memory performance, the negative health consequences of smoking outweigh any remote potential benefit. The good news for smokers is that it's never too late to quit, and the benefits of cessation are possible at any age. Although some people report a slight weight gain when they stop smoking, regular physical exercise can help offset weight gain and other physical and emotional responses to quitting. And imagine how nice it will be not to have to slip out of your nephew's wedding every 15 minutes to have a smoke, while Aunt Emma gives you the evil eye. Who needs that?

Alcohol: How Much Is Too Much?

For many people, drinking alcohol isn't really a lifestyle choice but merely a routine part of their daily social interaction. Having a work meeting over drinks has become as universally accepted as meeting for lunch. And, in some circles, even that lunch may involve two martinis. Alcohol is one of the most common substances that people both use and abuse. The health hazards of excess alcohol consumption are well known, from drunk-driving fatalities to liver disease. In terms of brain health, prolonged alcohol indulgence damages brain cells and leads to serious memory loss. And yet, surprisingly, studies have shown that some intake of alcohol, in moderation, may actually be good for our brains.

Sarah H. had been concerned about healthy living for most of her adult life: she was only an occasional social drinker, walked at least 20 minutes every day even after her husband and walking partner died, and rarely missed her daily crossword puzzle. For a 72-year-old, Sarah's memory was outstanding. But her identical twin sister, Lydia, took a different approach to life. Known as the "party girl" of the two sisters, she had been a heavy drinker and smoker for many years, and abhorred exercise. She laughed at the suggestion she might be an alcoholic, claiming she just "liked to have a good time." But as she got older, it seemed her body had more trouble tolerating the excesses. Several times during the past year, Sarah had found Lydia blacked out after one of her "parties."

Despite their identical genetic makeups, the twins had different cognitive abilities as well as memory capacities. Lydia was in the beginning stages of Alzheimer's disease. Clearly, the twins' differing lifestyle choices had contributed to the differences in their cognitive function as well as Lydia's developing Alzheimer's disease. For Lydia, drinking, smoking, and lack of exercise appeared to have played a major role.

Sarah and Lydia are not alone in their experience. Our own UCLA studies of twins confirm that genetic predisposition is only one determinant of risk for brain aging. Lifestyle choices to drink, smoke, and eat fatty diets can contribute to cognitive decline even in identical twins.

An eight-year study from Rotterdam, Holland, found that mild-to-moderate alcohol consumption—defined as one to four drinks each day—actually *lowered* a person's risk for developing severe memory loss. A similar study from Bordeaux, France, found that moderate wine drinkers had a lower risk for Alzheimer's disease. In fact, the risk for developing any kind of serious cognitive impairment was lower than for either heavy drinkers or non-drinkers. This type of moderate alcohol use has other health benefits, lowering the risk for heart attacks as much as 40 percent in one recent study.

In North America, moderate alcohol consumption is sometimes defined as up to two drinks for men and one drink for women, per day. Although this level of drinking could worsen one's risk for heart disease and stroke, the antioxidant effects of alcohol may slow down brain aging by interfering with free radical formation and inflammation. Exactly how alcohol might protect the brain or heart is not fully known, but it may involve an anti-platelet effect that lowers the blood's tendency to clot and cause tissue damage.

Many experts argue for red wine as the preferred brain fitness beverage because of its particularly potent antioxidant capacity. If someone does not drink alcohol, experts rarely recommend that they start drinking, because the potential hazards still outweigh any possible benefits. However, heavy drinkers should definitely cut back, and light to moderate drinkers need not quit to continue protecting their brains.

Memory Effects of Recreational Drugs

Many of today's baby boomers experimented with recreational drugs during the sixties, seventies, and beyond, and some used them regularly—particularly marijuana. LSD, amphetamines, and other hallucinogens were also popular, and in the late seventies and early eighties, cocaine became the drug of choice for many young, upwardly mobile people, or yuppies. As many users eventually became aware of the potentially harmful effects of recreational drugs, they gave them up.

I often hear questions and concerns regarding drug use in the past, and even decades ago. Studies on psychoactive drugs have shown they do affect memory abilities, but there are currently no data indicating exactly how long those effects last after a person stops using the drug.

Nick J., age 51, still lives and works in his northeast college town, where he owns a successful restaurant. He put himself through college by dealing marijuana, although he smoked up half his profits. After graduating, he went to chef's school and stopped dealing, but he kept up his pot habit. He had occasionally experimented with LSD, cocaine, and amphetamines, but marijuana was his drug of choice and had become part of his daily lifestyle. To Nick, grass was merely a social convention, like his parents having their cocktails before dinner. He didn't consider it to be a harmful or addictive drug—nothing like heroin or even alcohol.

Nick started to notice subtle memory changes about five years ago when he first began having trouble remembering the names of his regular customers. Then he started making errors in reservations, and even staff scheduling became a nightmare. His ex-wife actually accused him of "purposefully forgetting" to pick up their sons on two of his designated weekends.

When Nick sought professional help, his doctor did note some mild memory impairment, but his overall memory test scores were in the low normal range for his age group. The doctor strongly advised him to stop using marijuana, and explained that although Nick might experience some physical and emotional hurdles in giving up a decades-old habit, the evidence was overwhelming that chronic marijuana use worsens memory.

Quitting pot was harder than Nick had imagined. But after a few failed attempts, and with the support of his girlfriend and his business partner, he managed to get through the first three months pot-free. Nick noticed his memory improving after only a few weeks, and by month three others began to comment on his sharpness. He also started experiencing more energy, less moodiness, and an improved libido. His girlfriend was absolutely thrilled—she had always hated his moodiness.

Nick did continue to have mild memory complaints, but he no longer had difficulties that interfered with his job, and his memory test results remained in the normal range for his age.

Marijuana has been the most widely used illicit drug in many developed societies. Chronic marijuana use can impair memory, attention, and the ability to process information. Someone intoxicated from marijuana has a hard time recalling recent events and learning new information. Studies of chronic and heavy marijuana users show that they have difficulty with verbal and visual memory and attention. Despite the potentially harmful effects of marijuana, memory effects do diminish after people stop or cut down on their use.

A rash of new psychoactive drugs has emerged in recent years, including the popular drug known as Ecstasy. Animal studies have shown that Ecstasy causes damage to the brain cells that produce serotonin, the neurotransmitter that modulates mood and keeps us from becoming depressed. Extensive Ecstasy use has been shown to impair verbal and visual memory. A recent study found that even after a year of abstinence, Ecstasy users still showed evidence of memory impairment compared with those who had never used the drug.

In an atmosphere where new and untested drugs are being produced in bathtubs and widely distributed, people who experiment with these recreational drugs are putting their brain health in greater jeopardy than they know. At the risk of sounding like a square, my recommendation is to just say no.

Don't Overeat, Don't Overdrink, But Get Out and Make Merry

In addition to encouraging physical and mental activity, a major finding of the MacArthur Study of Successful Aging was that staying in close contact with the people in our lives, as well as remaining involved in sports, hobbies, charitable causes, or other meaningful activities, were key elements to success in aging. The more personally invested we are in a given activity, the more our ongoing health will benefit from it. Because a large component of MacArthur's definition for successful aging was cognitive success, such activities will likely promote brain health as well.

Remaining engaged with people means giving as well as getting support, and this support can take many forms. By maintaining close friendships, stable marriages, and long-term relationships, as well as spending time with people we love, respect, and esteem, our brains will function better in the long run. Research has linked healthy social relationships to greater longevity.

The practical support we get from close relationships may lead us to seek better medical care, and just hanging out with people who live a healthy lifestyle—like not smoking, or eating a low-fat diet—may rub off on us, too. Social support, or the emotional and practical advantages we gain from others, may even directly benefit us biologically. In a sense, we are "hard-wired," or genetically programmed, to interact with others. Talking, touching, and relating to others are key to maintaining well-being in our lives. It is within social groups that we protect each other and share our joys and concerns.

A recent study of men found that good social support significantly lowered their levels of epinephrine, norepinephrine, and cortisol—all physiological measures of stress. The evidence makes a strong argument for avoiding isolation and remaining engaged with others to keep mentally and physically healthy.

Juan R. was a jovial 72-year-old retired mechanic who had been active and relatively healthy all his life, except for his recently developed type 2 diabetes, which he controlled with diet and medication under the watchful eye of his wife, Carla. When Carla died suddenly from a stroke, Juan was distraught and nearly catatonic for a week. His daughter, Anna, was afraid to leave him alone in his apartment. Juan gradually got back on his feet, but he remained withdrawn and depressed, and Anna felt uncomfortable leaving him alone to cook and care for himself.

Anna knew from her mother's constant complaining that left to his own devices, Juan would sit in front of the television all day eating junk food, while his blood sugar ran all over the map. Since Carla's death, Juan hardly ate at all and had no interest in attending his weekly poker game with his buddies.

Anna already had her hands full with her own job and her kids. One afternoon, she arrived with Juan's groceries to find him passed out on the floor. The emergency room doctor diagnosed a hypoglycemic attack and clinical depression and recommended better diet supervision and a daily antidepressant for Juan.

That evening, after discussing it with her family, Anna insisted that Juan move in with them. It was no use protesting, and she would not take no for an answer. Besides, she persisted, she needed the free babysitting. He finally gave in, saying that Anna, like her mother, would never stop yapping until she got her way.

With her dad under the same roof, it was easier for Anna to keep an eye on him, and she started dropping him off at a local senior center three days a week. Juan enjoyed watching the kids and helping out with their homework, so Anna and her husband could actually get out some evenings.

Thanks to Anna's healthy cooking and her supervision of his medications, Juan's diabetes got back under control, his spirits picked up, and he eventually got back to beating his pals at poker.

Sex and Memory

We've seen how remaining engaged in meaningful pursuits, maintaining healthy relationships and physical activities, and making other smart lifestyle choices can go a long way to ensure physical and mental health as we age. Evidence also points to maintaining a meaningful sex life as yet another way to keep our brains young. People with healthy, active sex lives also tend to be more engaged in life and physically and mentally active—traits associated with successful brain aging. The physiological and mental benefits of a healthy sex life may well promote brain function as we age.

Systematic studies have shown that our sexual attitudes and interests are relatively consistent throughout life—surveys of older people indicate that nearly all of them would like to engage in sex if the time, place, and partner were right. The lack of available partners, however, is often a barrier to an active sex life for many older adults. Women live longer than men, and by age 85 there are twice as many women alive as men. Approximately 20 percent of men who live to age 65 or older have low testosterone levels and difficulties with ejaculation. The new drugs for male impotence, however, have clearly had an impact on this problem. Viagra is both safe and effective for many forms of erectile dysfunction for men even in their eighties or nineties, helping them to maintain a healthy sex life, which likely contributes to their brain health.

The physical illnesses some of us experience as we age, as well as the drugs we may take for these illnesses, can interfere with our sex lives; however, these challenges can be managed. For instance, patients with arthritis might schedule their analgesic medicine to kick in just prior to a romantic encounter, when increased flexibility might be desirable. Estrogen and testosterone replacement therapies can also help facilitate a healthy sex life when indicated, and current research is exploring these hormones' direct benefits on memory (Chapter 9).

Sleep On It

Sleep deprivation—another way we stress our brains—is a common problem for today's baby boomers, with their two-career families and multitasking lifestyles. Chronic insomnia and sleep deprivation can be either a symptom or a cause of depression, psychological stress, or both (Chapter 4). It can have a devastating effect on relationships and work performance.

Acute or chronic sleep deprivation impairs mental abilities and can lead to high blood sugar levels due to insulin resistance, a precursor of type 2 diabetes, as well as elevated brain cortisol levels—both associated with memory loss. Fortunately, this insulin resistance and memory impairment can be reversed if you spend approximately twelve hours in bed to make up for sleep debt. If the practical sleeping tips in Chapter 4 are not effective, professional help should be sought.

Aluminum on the Brain: Don't Toss Out Your Pots and Pans Just Yet

Nearly every time I give a lecture on Alzheimer's disease and memory loss, someone asks about the possibility that aluminum exposure contributes to the disease. People are concerned about using aluminum cooking utensils, deodorants, and a variety of aluminum-containing products.

Scientists have looked at this potential relationship in numerous ways. In a recently published eight-year study, French scientists found a twofold increase risk for Alzheimer's disease in geographic areas with higher aluminum concentrations in the drinking water, although these findings have not been replicated. Autopsy studies have detected some collections of aluminum in damaged areas of the brains of Alzheimer's patients, but the studies have not yet provided conclusive results. It is possible that aluminum collects in brain areas after the damage occurs, rather than actually causing the damage.

Other Lifestyle Pitfalls

Exposure to lead, pesticides, environmental mold, or any toxic chemical does have the potential to damage brain cells. Pesticide exposure in particular has been under recent scrutiny because of its possible influence on risk for Parkinson's disease. Although many of the epidemiological studies have not proved a direct connection between exposure to environmental toxins and the subsequent development of Alzheimer's disease, individual cases have been reported, and it seems logical to avoid chronic or acute exposure.

Making lifestyle choices is not just about our resolve to change but also becoming informed about the right kinds of changes to make. Dr. Vladamir Hachinski and his colleagues at the University of Western Ontario studied brain autopsies of people who died from various forms of dementia and found that those with lower educational achievement showed more evidence of small strokes in the brain than those who had attended college and further. An individual's educational achievement not only indicates their intellectual ability but also points to their ability to make better lifestyle choices that protect their brain. The study subjects with fewer years of education were more likely to smoke, eat fatty foods, and avoid exercise, activities that increase the risk of strokes and the likelihood of dementia later in life.

Making Lifestyle Choices to Keep Our Brains Young

Most of us know it's best to avoid head trauma, sleep deprivation, and smoking to maintain physical and mental health, and keep our memory ability at peak performance. The daily lifestyle choices we make have long-lasting effects. A professional boxer is bound to get hit in the head; a motorcyclist without a helmet is at risk to crack his skull. Many people choose more wisely, but that is not enough to actively keep our brains young. Avoiding nicotine and other toxins, keeping physically fit, moderating alcohol use, maintaining close and healthy personal relationships, and engaging in stimulating careers and pursuits are crucial to slowing brain aging and maintaining health. The Lifestyle Choices box summarizes how to maintain our health and protect our brains.

Lifestyle Choices That Keep Brains Fit

- Start an exercise program to maintain aerobic fitness, flexibility, and peak memory performance.
- Get both the aerobic and social benefits of walking with friends.
- Choose sports and physical activities with low risk for head trauma.
- Never drink and drive and always wear your seat belt.
- Wear helmets when riding bikes or doing sports.
- If you smoke, quit. Ask your doctor for help.
- If you drink alcohol, do it in moderation (up to two glasses of wine for men and one for women per day).
- Get involved in activities that have personal meaning. Spend time with friends and family.
- Get plenty of sleep.
- Avoid exposures to pesticides, organic solvents, molds, and other potential toxins.

Chapter Nine

Wise Up About Medicines

My doctor says too much sex can cause memory loss.
Now, what was I about to say?
—Milton Berle

In the early 1900s, most people could expect to live into their fifties—what we now consider middle age. Today, the average American man lives to age 73, while women can anticipate 79 years. Some experts estimate that by the year 2010, the average man will live to 85 and the average woman to 91.

Of the many advances that have helped to bring about this lifespan revolution, drug development is the leader. Although antibiotics, antiseptics, steroids, and other medicines have undoubtedly helped us to live longer, they have not necessarily helped us to live better. Only recently have drugs become available to treat memory loss and other cognitive conditions, or possibly prevent Alzheimer's disease.

Even armed with lifestyle strategies to keep our brains young, the most powerful tool for preventing mental decline as we age may come from new drug development. Wisely using currently available medicines now, if indicated, and adding new medicines as they are developed, is surely a key element in maintaining brain fitness and protecting against Alzheimer's disease.

At UCLA, we are testing cholinergic drugs—FDA-approved medicines known to boost memory and cognitive abilities in Alzheimer's disease victims—as "smart drugs" for people with normal memory. Researchers are also discovering unexpected memory benefits from medications currently marketed for other conditions, such as anti-inflammatory drugs and estrogen replacement hormones. Scientists have begun human testing on a vaccine that may not only prevent the brain from accumulating Alzheimer's plaques but may also eliminate already existing plaques.

How Doctors Understand Medical Conditions

The first step to using medicines wisely is to communicate effectively with your doctor, the person who will write the prescription and advise you on how to use the medication. With managed care and other pressures in the current health care system, physicians don't always have time to sit about and chat with their patients, so a concise, focused approach is your best bet. One of the more useful lessons I learned in medical school was the way in which doctors gather and organize information about their patients. Once I learned this straightforward system, I found that visits to my own doctor became more efficient. He appreciated my organized description of symptoms, and the approach helped us both remember to cover important areas in my medical history.

When reviewing your health history, keep in mind some of the problems that can cause memory loss, such as depression, vision or hearing problems, infections, or poor nutrition. It's helpful to write a list of your symptoms before your doctor appointment. Also, try to be specific—provide details about the timing of symptoms, their quality, and any events associated with their onset. Bring in all your medicines, or a detailed list, to avoid confusion about what pills and how many of them you are taking. Ask questions, request explanations, and try to answer the doctor's inquiries as honestly and accurately as you can. Your physician should explain the diagnosis and the pros and cons of alternative treatments. Therapeutic options often include both medications and non-medicinal approaches.

A physician's evaluation of memory loss usually involves a thorough history, a physical, neurological, and mental status examination, and a laboratory assessment. In the mental status exam, the doctor will screen for depression, memory loss, and other cognitive difficulties. Usually a brief mental status exam can be completed within 10 to 15 minutes, but more detailed memory assessments, or neuropsychological tests, can provide a better understanding of subtle memory deficits. The laboratory component should include blood tests to rule out thyroid disease, vitamin B_{12} deficiency, and other disorders that could possibly cause memory changes. And, as noted in Chapter 1, a PET scan is the most sensitive method of detecting possible Alzheimer's disease.

If you are planning to see your physician for a memory loss assessment, I suggest preparing yourself by considering the format doctors use in gathering and organizing information about their patients (see box). It covers many key points of relevance to brain aging and memory loss and could help you form a partnership with your doctor and become proactive in your medical care.

How Doctors Gather and Organize Information About Patients

Identifying data. Brief description of the patient (age, race, marital status, etc.).

Chief complaint. The reason the patient is seeing the doctor.

History of present illness. The nature, onset, and progression of memory symptoms. Description of other relevant problems, such as depression, anxiety, and stress; time course and events related to the symptoms.

Past medical history. Other potentially pertinent physical conditions, including hypertension, diabetes, prior head trauma, increased blood cholesterol, Parkinson's disease, strokes.

Past mental history. Previous depression, memory losses, and other relevant details are recorded along with treatments received.

Family history. Parents, siblings or other relatives who had Alzheimer's disease, other dementias, or any of the above-mentioned medical or mental illnesses.

Medications. Current and relevant past medicines used, including over-the-counter drugs and supplements, with an emphasis on medicines that affect memory.

Social/personal history. A record of education, work, marital, and other relevant social history, as well as lifestyle choices and potential risks and protections for memory loss. Dietary habits, drinking, and smoking patterns are recorded.

Mental status. Includes assessment of appearance, behavior, memory, orientation, mood, judgment, and insight.

Physical examination. Findings from the examination, ranging from blood pressure and pulse to observed physical signs and abnormalities.

Laboratory assessments. Results of screening blood tests, brain scans, and other laboratory findings.

Impression. A summary of the most likely diagnoses and problems.

Plan. A listing of specific interventions for each of the above problems.

Getting Treatment for Physical Illnesses

Like all the other aging boomers, I am at risk for developing a variety of physical illnesses, including hypertension, high cholesterol, and diabetes, that can impair memory ability and affect long-term brain health. Studies have found higher rates of Alzheimer's disease and other dementias in people with these conditions. Effective medicines are readily available to treat these illnesses, and getting appropriate and timely treatment is vital to keeping our brains young.

I have friends who have gone to their doctors for a head cold or a sprained ankle, and during the exam the doctor discovered an elevated blood pressure or a high blood sugar level. Picking up these incidental findings can not only save a person's brain cells but their life as well. This is just one of many reasons why regular physical checkups are so important.

Hypertension

High blood pressure, or hypertension, affects more than 60 percent of people over age 65. The illness packs the added punch of increasing a person's risk for strokes and vascular dementia, as well as heart attacks. Hypertension has been described as a silent epidemic because most of us wouldn't know we had it unless we had our blood pressure measured. High blood pressure can be easily and effectively treated with a variety of proven medicines, but the most effective intervention for hypertension usually includes both medicine and lifestyle changes. Smoking, overuse of alcohol, and being overweight all contribute to one's hypertensive risk. Regular exercise, a low-salt diet, and avoiding smoking and other high-risk activities all lower blood pressure.

Recent research shows that chronic high blood pressure during midlife (forties and fifties) leads to cognitive decline later in life.

Chronic hypertension most likely affects memory because it thickens and stiffens blood vessels. Under high pressure, these stiffened blood vessels can rupture, possibly causing cerebral vascular disease involving blood leakage into the brain tissue and stroke. A stroke is often defined as the death of brain cells, resulting in a loss of physical or mental function or both. But treatment makes a difference. Dr. Edwin Jacobson of UCLA recently reported that rigorous control of mild to moderate hypertension can improve cognitive function. He noted significant improvements in visual and spatial skills, executive skills, and the speed that patients could process information after just twenty-four weeks of treatment.

Many Alzheimer's patients also show evidence of cerebral vascular disease, which can further compromise their cognitive status. A recent autopsy study of patients who had been diagnosed with Alzheimer's disease found that roughly one-third had cerebral vascular disease as well. There are also some patients with vascular brain injuries caused by hypertension and other illnesses who also have Alzheimer's plaques and tangles in their brains. The coexistence of Alzheimer's and cerebral vascular disease is much worse than experiencing either alone.

The Ups and Downs of Cholesterol

High blood cholesterol increases the risk for strokes and other circulatory problems that can affect memory. Inherited genetic factors as well as lifestyle choices are known to contribute to the risk for high cholesterol. In recent years, the class of cholesterol-lowering drugs known as *statins* have been found not only to lower fat levels in the blood but also to help prevent age-related memory decline. These cholesterol-lowering drugs are known to prevent heart disease and stroke, and new research indicates that people who take them also have a lower risk for developing Alzheimer's disease.

Dr. Benjamin Wolozin and Dr. George Siegel and their colleagues at Loyola University studied more than 60,000 hospital medical records. They found that the rate of Alzheimer's disease in patients taking cholesterol-lowering statins, including lovastatin (Mevacor) and pravastatin (Pravachol), was nearly 75 percent lower when compared to the entire population, or to patients taking other medicines for different conditions such as hypertension or cardiovascular disease.

Dr. David Drachman, University of Massachusetts, found that a wide variety of statins have the effect of lowering one's risk for Alzheimer's, including atorvastatin (Lipitor), cerivastatin (Baycol), fluvastatin (Lescol), pravastatin (Pravachol), and simvastatin (Zocor). Scientists speculate that when the statin drugs interfere with cholesterol metabolism, they may also decrease the production of amyloid-beta, which forms the Alzheimer's plaques. The benefits to the brain from statins may also stem from their ability to reduce cerebral vascular disease, thereby improving blood circulation to brain cells. Until a double-blind test comparing statin drugs against a placebo control is completed, we cannot state definitively that these drugs truly help to prevent Alzheimer's disease. However, current data are encouraging. In addition, the mounting scientific evidence of the cardiac benefits of cholesterol-lowering drugs recently spurred the National Heart, Lung and Blood Institute to change their guidelines to recommend that a greater number of Americans should be taking these drugs—about 36 million compared to the 13 million of previous guidelines.

Coronary Bypass Surgery

High cholesterol, high blood pressure, and other conditions sometimes damage the heart to the extent that surgery is needed. A recent report published in the *New England Journal of Medicine* noted a startling decline in memory ability in people five years after they had undergone coronary-artery bypass surgery. They found that memory and other cognitive declines were present in 53 percent of the patients at the time of discharge from the hospital after their bypass surgery. This high rate went down by about half after six months. However, five years later, the proportion of cognitively impaired patients was back up to 42 percent. The main predictor of memory loss five years after bypass surgery appeared to be lower cognitive function at initial hospital discharge following the surgery.

To perform coronary bypass surgery, the doctor has to stop the patient's heart and divert their blood through the artificial pump of a heart-lung machine. Some doctors believe that during this stopped-heart period the brain sustains subtle damage. Another theory is that the operation may shake loose tiny particles of fat from the surgical site into the blood system. If and when these droplets make their way to the brain, they can cause cellular damage. The heart-lung machine also produces air bubbles that could block blood flow through tiny vessels, thus killing brain cells. The *New England Journal of Medicine* report will likely open the door for increased study of prophylactic treatments to prevent post-surgical cognitive decline, including the use of anti-Alzheimer's drugs.

Anesthesia used during surgery is also being studied as a possible contributor to post-surgical memory decline. Although most people have no long-term cognitive losses after one or two surgeries, the cumulative effect of multiple exposures to anesthesia has the potential to accelerate brain aging, particularly in someone already at risk. A recent study of people over age 64 found cognitive decline in 53 percent of them up to three months after surgery. Several studies have found a greater susceptibility to dementia in older people who undergo surgery for hip fracture. It is possible that the decrease in blood pressure from anesthesia lowers blood circulation in the brain, resulting in neuronal death in vulnerable brain areas.

Because of the general risks of surgery and the possible risks to brain health from cumulative exposures to anesthesia, you may want to consider whether elective or non-crucial surgery is the best lifestyle choice for you.

Other Physical Illnesses

Diabetes, a disease resulting from the body's inability to adequately control sugar levels in the blood, is another illness that can impair memory and brain fitness, and becomes more frequent with age. An estimated 16 million people in the United States suffer from diabetes, yet about half of them don't even know they have it. In addition to exercise and diet (Chapter 7), there are well-known medications, like insulin, that effectively treat diabetes. Control of diabetes will protect brain fitness and can improve memory function.

Any acute illness that attacks our bodies can overwhelm our brains (see Table 9.1). I have seen both older and younger patients experience memory impairment and word-finding difficulties during a flu or pneumonia. Usually these problems disappear or are reversible when the disease lifts. One common mistake many people make is to discontinue taking their antibiotics once they begin to feel better. A full course of antibiotics is crucial to prevent an infection from recurring, as well as preventing their bodies from building a possible immunity to that medicine.

The bottom line is to take your physical illnesses seriously, seek medical advice when indicated, and use medicines wisely.

Table 9.1

PHYSICAL CONDITIONS THAT MAY CAUSE MEMORY LOSS

Cardiac Disorders		
	Arrhythmias	Heart attack
	Congestive heart failure	
Infections		
	Encephalitis	Meningitis
	Hepatitis	Pneumonia
	Influenza	Tuberculosis
Malignancies		
	Brain	Lymphoma
	Leukemia	Pancreas
Metabolic and Endocrine Disorders		
	Cushing's disease	Electrolyte imbalance
	Dehydration	Thyroid disorders
Neurological Disorders		
	Epilepsy	Parkinson's disease
	Multiple sclerosis	Strokes
Other Conditions		
	Anemia	Pain
	Kidney disease	Temporal arteritis
	Liver disease	Thiamine deficiency
	Lupus	Vitamin B_{12} deficiency

Drugs That Can Worsen Memory

As our bodies get older, they become more vulnerable to certain physical illnesses and we tend to take more medicines. The average older adult takes more than half a dozen prescription medicines at any one time. The more medicines we take, the greater the possibility for negative drug interactions.

Aging causes our brain receptors to become more sensitive to the effects of medication, and this sensitivity can lead to side effects at much lower doses. Also, our bodies become less efficient in metabolizing and excreting medicines, so over time we may accumulate higher blood levels of these drugs. This can lead to new or increased side effects, as well as to interactions with other drugs that we hadn't experienced in the past. Due to these changes in our bodies, doctors caring for older people often prescribe drugs in low doses initially and slowly increase them as needed to minimize any potential adverse reactions.

Many drugs have anticholinergic side effects, making them oppose the actions of the drugs prescribed for memory loss, thereby worsening memory ability. Drugs of particular concern include those often prescribed for anxiety—such as Xanax, Valium, or Librium—which can cause sedation and can also impair memory ability (see Table 9.2). Drugs used to regulate heart rate or treat high blood pressure can make blood vessels less taut and decrease the heart's ability to pump blood. Since our vascular tone diminishes anyway as we age, medications that aggravate this problem can lead to falls, head trauma, and other complications that threaten brain fitness, and they should be taken with care.

When a patient visits the UCLA Memory Clinic for the first time, we ask them to bring in all their medications. Many older patients have arrived with shopping bags filled with prescription bot tles. Often the easiest and most effective interventions are merely to eliminate unnecessary medicines and reduce the dosage level of drugs that are most likely contributing to the patient's memory loss, depression, or both. Giving a patient the lowest effective dose of a medication will lower their risk for developing side effects; however, too low a dose, or a sub-therapeutic level, can also be a problem.

Doris L., widowed for seven years, lived in a beautiful Park Avenue penthouse near Central Park South. Her successful surgeon husband, Melvin, had always treated the family's minor ailments, prescribing antibiotics, ointments, or eye drops whenever Doris, the kids, or the grandkids needed anything. After Melvin died, Doris urgently shopped for a physician to fill the void of not having Melvin there to soothe her every ache and pain. Though several physicians throughout Manhattan had treated her once or twice, she couldn't bring herself to trust any one doctor to truly understand her "complicated medical conditions," which included high cholesterol, arthritis, chronic insomnia, migraines, and periodic depression. Doris took a different medicine for each ailment.

Throughout her adult life, Doris had experienced bouts of depression that recurred every two to three years. Melvin had always prescribed an antidepressant, which seemed to help. After his death, she experienced an extreme episode of depression and asked one of her new physicians to continue the antidepressant prescription. Her depression improved, but she began to complain of increasing forgetfulness. At first she merely misplaced things, but her symptoms worsened. She would put on her coat, get her purse, and push the call button for the elevator. When the doors opened, half the time she would have already forgotten where she was going.

Doris consulted a new doctor, who suspected she was experiencing some early symptoms of Alzheimer's disease and prescribed a cholinergic drug. After a month, she noticed improved memory ability. A week later, Doris read about a newer anti-Alzheimer's drug that sounded fantastic. Concerned about insulting her current doctor, she sought yet another physician to prescribe this new drug. Doris was now taking a daily antidepressant, a sleeping pill, and two cholinergic drugs, not to mention meds for her high blood pressure, arthritis, high cholesterol, and headaches. The second, unnecessary, cholinergic, which neither prescribing physician was aware of—Doris made sure of that—led to nausea and vomiting, which forced her to consult a gastroenterologist. This specialist insisted on seeing all her medical records and all medicines before he could help her.

Desperate, Doris came clean. Apparently, she was being treated by seven different doctors and getting multiple prescriptions for many of her conditions including pain, anxiety, depression, and memory loss. Most of her symptoms were intensified by the overuse and interactions of her drugs. The antidepressant she was taking actually had an anti-cholinergic effect that worsened her age-related memory loss. Also, the doubling up on the cholinergic drugs only served to upset her stomach, but failed to give her any extra memory improvement over the original single dose. Her insomnia was a symptom of depression, so rather than a sleeping pill—which also worsened her memory—her gastroenterologist prescribed a newer antidepressant that helped with her sleep and had fewer side effects. He also reduced the pain from her arthritis and headaches by using a nonnarcotic anti-inflammatory drug.

Although her depression and memory function improved, Doris still missed Melvin. She began seeing a psychotherapist, who helped her to realize that her "doctor shopping" had been an ineffective attempt to feel cared for—the way she felt before Melvin passed away. That would never be possible. Doris began to accept things she couldn't change and move on with her life. She became involved in social and volunteer activities and actually started dating two years later—a nice dentist from midtown.

If you are concerned that medication may be affecting your memory, you should consult your doctor about whether you truly need a particular drug, and be sure she is aware of all the medicines you take. This is especially important if you are under the care of more than one physician.

Table 9.2

COMMON MEDICINES THAT CAN IMPAIR MEMORY IF NOT TAKEN WISELY		
DRUG CATEGORY	COMMON NAME	GENERIC NAME
Anti-ulcer/ dyspepsia	Tagamet Zantac	Cimetidine Ranitidine
Analgesic	Codeine Demerol Fiorinal Percodan, Percocet	 Meperidine Oxycodone

DRUG CATEGORY	COMMON NAME	GENERIC NAME
Antihistamine	Benadryl	Diphenhydramine
Anti-anxiety/ Sedative	Ambien	Zolpidem
	Ativan	Lorazepam
	Dalmane	Flurazepam
	Halcion	Triazolam
	Librium	Chlordiazepoxide
	Restoril	Temazepam
	Serax	Oxazepam
	Valium	Diazepam
	Xanax	Alprazolam
Antidepressant	Elavil	Amitriptyline
	Tofranil	Imipramine
Anti-hypertensive	Benazepril	Lotensin
	Dyazide	Hydrochlorothiazide
	Tenormin	Atenolol
	Toprol	Metoprolol
Anti-Parkinson's	Symmetrel	Amantadine
Anti-psychotic	Haldol	Haloperidol
	Mellaril	Thioridazine
	Thorazine	Chlorpromazine
Hormone	Synthroid, Thyroxine	Thyroid supplements
Steroid	Prednisone	Prednisone

The Business of Anti-Aging

Consumers spend billions of dollars a year on "cures" for the aging process. Over-the-counter treatments, including ginkgo biloba, ginseng, and melatonin, as well as various vitamins and herbs, are proclaimed to be the answer we've all been seeking—veritable fountains of youth. However, despite dramatic claims of their effectiveness, these treatments and herbal tonics are not always rigorously tested, nor are they monitored by the FDA. It is fairly easy to make grandiose claims about unproven supplements—often touted as memory loss preventions in an unregulated, multibillion-dollar industry.

The power of the "placebo effect" may help explain the popularity of these treatments. I know one mother who used this tried-and-true placebo effect to treat her children's ailments. When one of them had a mild ache or pain, she would go to her medicine cabinet and pull out a large bottle of cherry-flavored liquid labeled "PLACEBO." She would then dole out a large tablespoon of the syrup while commenting on its remarkable potency. The placebo usually reduced and sometimes even cured their ailments.

Scientists have long speculated on what causes this scientifically proven placebo effect—whether it is the patient's belief or expectation that the treatment will work, the doctor's confidence in the treatment, or some other physiological mechanism. In memory studies we have found that sugar pills often make us better, but this improvement is temporary, usually lasting no more than six weeks.

As a result of the diminishing placebo effect, the gold standard for proving the effectiveness of any new drug or treatment is to show that the active drug is indeed more beneficial than a placebo—over time.

A recent study analyzing 114 published reports involving 7,500 patients with a variety of different conditions raised new controversy about the placebo effect. Dr. Asbjorn Hrobjartson and Dr. Peter Gotzsche of the University of Copenhagen found that what many believe to be an effect of placebo is merely the result of the natural uneven course of an illness. When treatments were compared with no treatments, study participants not receiving treatment improved at about the same rate as when participants were given a placebo. Despite these results, placebos are still needed in clinical research to prevent scientists from knowing who is getting the real treatment, and who is not.

Drugs used to treat Alzheimer's disease or prevent memory loss must undergo intense placebo-controlled testing before gaining FDA approval for distribution. In Alzheimer's disease treatment studies, we have often seen an initial but temporary placebo effect, as illustrated in Figure 9.1.

Figure 9.1

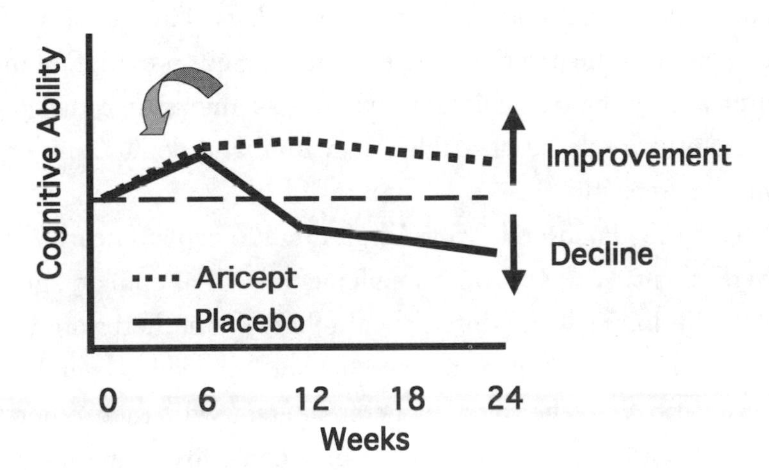

In the study graphed in the figure, hundreds of patients with Alzheimer's disease were given either the active cholinergic drug donepezil (Aricept) or an identical-appearing placebo pill. Note the gray arrow pointing to the first six weeks of the study—the placebo was just as good as the Aricept in improving memory and other cognitive abilities. This effect, however, waned after that initial period and the placebo patients began to decline, while the patients taking Aricept maintained their cognitive improvement. In fact, other studies indicate that this drug continues its benefit for several years. People who take treatments that have not passed a controlled study are probably wasting their time and money (they could just as easily gain placebo effect from a breath mint) and may be causing themselves unnecessary side effects that can harm their health.

All That Is Natural Is Not Necessarily Safe

An estimated 124 million Americans have tried herbs, vitamins, minerals, enzymes, and other "natural" remedies to treat a variety of ailments. In the United States alone, over $30 billion is spent each year on these supplements, and more than 80 percent of the world population uses botanical preparations as medicines. The majority of us are unaware of the potential dangers of mixing our prescription medicines with herbal remedies, and we mistakenly assume that these treatments are safe because they are natural and don't require government oversight.

The U.S. Food and Drug Administration reports nearly 3,000 adverse events each year from supplements such as ephedra, ginkgo biloba, St. John's wort, ginseng, and others. Some herbs are therapeutic at a low dose but toxic at another and can lead to liver disease or even cancer. When various herbs interact with each other, let alone prescription drugs, they also can become toxic, interfere with the action of the drugs, or both.

Food manufacturers have been trying to build on their earlier successes with dietary supplements by adding herbs to foods. Because of concern about the potential harmful results, the FDA recently notified several companies that this practice violates federal regulations governing what manufacturers can and cannot add to food.

Of all the various herbal remedies marketed to treat the maladies of aging, ginkgo biloba has received the greatest attention in the last few years. An estimated 11 million Americans have used this 4,000-year-old herbal medicine, which is made from a leaf extract and thought to improve memory ability by inhibiting oxidative cell damage and improving cerebral circulation.

Ginkgo has been tested in several forms of memory impairment, including mild age-related decline, vascular dementia, and Alzheimer's disease. Some studies have shown significant results, but the clinical relevance of the effects has been unclear. Moreover, experts have questioned the methods of these studies. Improved and better-designed studies are currently under way, including a large trial comparing ginkgo biloba to placebo in approximately 3,000 people aged 75 or older, so we look forward to more reliable information in the future.

Because the FDA does not regulate ginkgo biloba, the quality and consistency of the many brands available differ considerably. Ginkgo has been known to cause nausea, heartburn, headaches, dizziness, excessive bruising or bleeding, low blood pressure, and other adverse effects. Ingesting ginkgo along with caffeine can lead to blood clots surrounding the brain (subdural hematomas), and ginkgo can affect insulin secretion, making it potentially dangerous for diabetics. Because ginkgo biloba has anti-coagulant properties, taking it with aspirin and other blood-thinning drugs requires careful monitoring. The current limited evidence for ginkgo's effectiveness, and its potential for adverse effects, leads me to recommend against ginkgo biloba as a treatment for preventing memory loss at this time.

Ephedra is often used as a stimulant or appetite suppressant, or as an ingredient in other medicines, usually over-the-counter preparations. If people combine ephedra with coffee or other mild stimulants, they may experience rapid heart rate and anxiety. Guarana is an herb used as a stimulant and natural source of caffeine, which has some anti-coagulant effects. Serious heart problems can result if it is mixed with ephedra. Ginseng, a popular herb used to increase endurance, reduce stress, and improve sexual function, can be harmful for diabetics. Ginseng supplements are sometimes packaged as orange-flavored treats and appear on the candy shelf rather than the supplement counter. Taken like mints or chewing gum, it is alarmingly easy for harmful doses to get into a diabetic person's body. If taken in conjunction with antidepressants, ginseng can make some people manic. It can also augment the effects of sedatives or stimulants.

Kava kava has been used to reduce anxiety and stress. It can enhance the effects of alcohol and lead to intoxication. In conjunction with sedatives, it can cause excessive sleepiness or even coma. Valerian, an herb used for restlessness and insomnia, can interact adversely with sedatives or alcohol. Another herb native to Europe, St. John's wort, is taken by more than 7 million Americans as a natural treatment for depression or insomnia. Although recent studies have not found it to be effective in clinical depression, it can augment the effects of antidepressants, stimulants, or anticonvulsants in some situations. Rarely, when exposed to sunlight, people using St. John's wort have experienced sensations of tingling, needle pricks, or pain, even when taking it alone.

The majority of people who use these natural untested remedies experience no negative side effects, and many individuals swear to their beneficial effects. In fact, herbal remedies may be effective for some people in certain situations that experts are not even aware of.

The scientific community and the public at large continue to wait for results of conclusive double-blind studies showing whether or not ginkgo biloba and other herbal memory loss treatments actually work. For now, considering their potentially harmful side effects, I usually recommend my patients avoid taking unnecessary risks with unproven anti–brain-aging remedies.

Pharmacogenetics: The Future of Alzheimer's Disease Prevention

In recent years, several available drug treatments have proven effective for the memory loss and other cognitive declines of Alzheimer's disease. But as scientists continually search to find a cure for Alzheimer's disease and possibly prevent its onset in the first place, a new approach is emerging: the science of *pharmacogenetics*—the strategy of treating patients not just according to their diagnosed disease but also according to their genetic makeup.

Particular genetic variants may lead to differences in an individual's drug response, and these differences can take effect at many levels. It could affect how well the drug is absorbed from the stomach and intestines, how well it is broken down in the body, how quickly it gets through the blood system and out of the body, or how effectively it binds to neurotransmitter receptors in the brain. By identifying and understanding the functions of these genetic variations, we can predict their role when determining a patient's response to a particular drug.

The goal of pharmacogenetics is to provide doctors with a patient's drug response profile before beginning medication treatment. A meaningful pharmacogenetic profile may help to define a sub-population of individuals who are likely to respond or not respond to a particular drug, based upon that population's underlying biology. The point is to avoid unnecessarily treating patients whose genetics indicate they may receive little benefit from a given drug yet have a high risk for side effects.

Because the damage is done by the time a patient develops Alzheimer's disease symptoms, I am convinced that our best chance of "curing" the disease is to target mild forgetfulness. At UCLA we are conducting a double-blind study to develop a pharmacogenetic treatment strategy for delaying the onset of Alzheimer's disease

(Chapter 1). All research subjects are tested for any APOE-4 Alzheimer's genetic risk. They also receive a PET scan before starting treatment, as well as a follow-up scan two years later. We predict that the volunteers who take placebo pills will show more rapid decline in brain function (i.e., accelerated brain aging) than those taking an active drug. We expect the actual drug to be more effective in volunteers with the APOE-4 genetic risk than those without the genetic risk. Our aim is to delay brain aging by one or more decades, thus allowing people to live longer, and better, without memory decline. Our hope is that these ongoing studies will lead to the widespread use of pharmacogenetic testing to determine who is or is not likely to benefit from Alzheimer's prevention drugs.

Maintaining Memory with Anti-Alzheimer's Drugs

New research on neurotransmitters suggests that early intervention with a cholinergic drug can slow brain aging as well as delay the onset of Alzheimer's disease. These studies support the idea that treating early brain-aging symptoms with cholinergic drugs—medicines currently used to treat Alzheimer's disease—may actually interfere with the deposition of amyloids, the insoluble proteins that have accumulated in the brains of people with Alzheimer's disease.

Dr. Diana Woodruff-Pak and her associates at Temple University treated rabbits with galantamine (Reminyl), a cholinergic drug that increases acetylcholine neurotransmitters. The neurotransmitter acetylcholine communicates with the brain's nicotinic receptors, which gradually decline as we age, and more rapidly if one has Alzheimer's disease. The researchers found improved learning abilities in both the young and old rabbits, as well as increases in nicotinic receptor activity in their brains.

The fact that the *young* animals showed improved learning has led some experts to speculate that the drug has an effect beyond its usual inhibition of the enzyme that breaks down acetylcholine. Even though cholinesterase inhibitor drugs were designed to increase the acetylcholine neurotransmitter directly, they have other effects that may slow down brain aging. Large-scale human studies are already under way in people with mild memory complaints to see if using such drugs early in the course of brain aging might modify the progression of cognitive decline.

Antidepressant Medicines: Good for Mood and for Memory

People become depressed for a variety of reasons. Often a personal loss or disappointment can trigger sadness. Unresolved conflicts or persistent stress are sometimes the culprits. Other times a biochemical imbalance in the brain may be to blame. Many depressions involve more than one trigger or cause, with overlapping psychological and biological factors contributing. Regardless of the specific cause, psychotherapy, antidepressant drugs, or both, can improve the symptoms, even if they are severe.

One feature of depression—decreased ability to concentrate—seems to become more prominent as we age. Middle-aged and older people tend to emphasize these concentration difficulties, and their depressions are often colored by memory complaints. A form of depression often seen in older people has been labeled as "pseudodementia" because it so closely resembles a dementia or Alzheimer's disease. A person who is overwhelmed by sadness and despair may have trouble trying to learn, and remembering new information is the last thing on their minds. Their sleep patterns are also disturbed, further aggravating memory abilities (see Recognizing Features of Major Depression box). It is known that episodes of repeated and severe depression can lead to abnormal secretion of stress hormones, and this can further worsen memory problems (Chapter 4).

Many people still consider depression to be a sign of character weakness. To them, seeking professional help or taking antidepressants is a stigma to be avoided at all costs. What those people don't realize is that untreated depression can increase a person's risk for serious physical illness or even death, as well as raise the risk for suicide. The mortality rate for patients whose depressions are properly treated is half that of those who receive inadequate care or no care.

Recognizing Features of Major Depression: "SIG E CAPS"

Many psychiatrists use a simple mnemonic to remember eight of the features of major depression. The memory tool stands for "when to prescribe the energy capsules": "SIG" is an abbreviation doctors use to stand for prescribe (check your next prescription from your doctor), "E" stands for energy, and "CAPS" stands for capsules. Each letter is an abbreviation for one of the symptoms:

S—Sleep decreased or increased
I—Interest decreased
G—Guilt feelings
E—Energy decreased
C—Concentration abilities decreased
A—Appetite decreased or increased
P—"Psychomotor" disturbance (pacing, hand-wringing, or slowing of thought and movement)
S—Suicidal thinking

If you or someone you know has four of these symptoms for two or more weeks, you should consult a doctor, because an antidepressant medicine is likely to improve the symptoms.

Geriatric psychiatrists have studied the combined states of memory loss and depression in older adults. Both Dr. George Alexopoulos at Cornell University and Dr. D. P. Devanand at Columbia University have found that the combined states, though treatable with antidepressants, still tend to progress to permanent cognitive losses that are characteristic of Alzheimer's disease.

Some experts wonder whether aggressive treatment of depressive symptoms might stave off the chronic memory loss and/or the dementia that eventually develops in many of these patients. The possibility that antidepressant drugs could slow down or prevent brain aging requires further study. However, we do know that both antidepressant medicines and psychotherapy are effective treatments in their own rights, but combining the two, when indicated, appears to be more effective than either form of treatment alone.

Although all types of antidepressant drugs have been effective in relieving some symptoms of depression, the improved side-effect profiles of the newer antidepressants, like fluoxetine (Prozac), sertraline (Zoloft), mirtazepine (Remeron), or citalopram (Celexa), to mention a few (see box), have become preferred treatments over older medicines like amitriptyline (Elavil) or imipramine (Tofranil). These older medicines can potentially worsen memory performance because of their anti-cholinergic side effects.

It is best to start low and go slow with antidepressant medicines, particularly when treating older people. Many primary care physicians can treat depression quite effectively using antidepressants, but for a complicated and more severe depression, the expertise of a psychiatrist may be needed. For some older depressed patients, a psychiatrist with additional geriatric training would certainly offer the most sophisticated care (Appendix 5).

Some Commonly Used Antidepressant Drugs		
DRUG	DOSE RANGE (MG/DAY)	POSSIBLE SIDE EFFECTS
Buproprion (Wellbutrin)	37.5–450	Insomnia, nausea
Citalopram (Celexa)	10–60	Insomnia, nausea, sexual dysfunction
Desipramine	10–300	Heart arrhythmia, sedation, sexual dysfunction
Paroxetine (Paxil)	10–60	Nausea, sexual dysfunction
Fluoxetine (Prozac)	10–80	Insomnia, nausea, sexual dysfunction
Mirtazepine (Remeron)	15–45	Sedation, weight gain
Nefazadone (Serzone)	100–600	Sedation
Nortriptyline	10–200	Heart arrhythmia, sedation, sexual dysfunction
Sertraline (Zoloft)	25–200	Headache, nausea, sexual dysfunction
Venlafaxine (Effexor)	37.5–375	Insomnia, nausea
Trazodone (Desyrel)	25–600	Sedation, heart arrhythmia

Using Sex Hormones to Prevent Memory Loss

In recent years, scientists have taken increasing interest in the effects of estrogen and other hormones on mood and memory in older adults. Studies by epidemiologists have found that taking estrogen supplements after menopause lowers a woman's risk for developing Alzheimer's disease. It is important to note that women who take hormone replacements are, on average, better educated and live healthier lifestyles.

Estrogen appears to improve connections between the brain's nerve cells and to augment cerebral blood flow. It also boosts memory transmitters like acetylcholine. In addition, estrogen serves as an antioxidant, helping to prevent damage to cells over time. Based on estrogen's proven benefits for treating menopausal symptoms like hot flashes and insomnia, hormone replacement therapy is already a $5 billion annual industry and continues to grow.

There is still much debate over estrogen's true benefits as well as its drawbacks. In addition to alleviating menopausal symptoms, estrogen may improve skin tone, prevent osteoporosis, and reduce the risk for stroke. However, estrogen taken without progesterone increases the risk for developing endometrial cancer and may increase the susceptibility to breast cancer. Use of estrogen replacement with or without progesterone is associated with a twofold increase in risk of developing gallstones.

Although evidence suggests that estrogen protects against some forms of age-related memory loss and may even help prevent Alzheimer's disease from developing in healthy brains, it has shown no apparent benefits in patients who already have Alzheimer's disease.

Dr. Susan Resnick at the National Institute on Aging found that women taking estrogen perform better on certain memory tests and have better blood flow to the hippocampus, a brain area involved in memory function. Dr. Barbara Sherwin at McGill University in

Montreal has shown that estrogen's benefit for post-menopausal women is concentrated on verbal, rather than visual, memory.

Hopefully, in the next few years the Women's Health Initiative will give us a definitive answer on whether or not post-menopausal estrogen use can truly stave off Alzheimer's disease. In this large-scale study, U.S. investigators are randomizing nearly 10,000 women to different estrogen preparations or placebo and following their rates of osteoporosis, heart disease, cancer, and Alzheimer's disease. If estrogen does turn out to protect women from developing Alzheimer's disease, then that benefit also needs to offset such potential risks as breast cancer and heart disease.

New drug development has led to synthetic estrogens designed to isolate specific beneficial effects while eliminating unwanted side effects. Although investigations thus far have not shown these "designer" estrogens, or selective estrogen receptor modulators (SERMs), to improve cognitive function, many experts remain optimistic of their potential benefit.

The male sex hormone testosterone also has some important effects on mood and memory. Men experience a drop in this sex hormone as they age, but instead of a rapid decline, the levels decrease very gradually over decades. Only about one out of five men 65 and older end up with an abnormally low testosterone level, and initial studies indicate that men with low levels experience improvements in mood and memory following testosterone administration. Our UCLA research group is among those systematically studying testosterone's potential benefits for memory.

Anti-Inflammatory Drugs—for Pain, Memory Loss, or Both?

In the late 1990s, epidemiologists found that using the anti-inflammatory drugs ibuprofen (Advil, Motrin, Nuprin), naproxen sodium (Aleve), and indomethacin (Indocin) was associated with reducing the risk of Alzheimer's disease by as much as 60 percent—if people took them for at least two years. Some scientists trying to explain the connection have theorized that the drugs' brain effects came from their action on inflammation. Looking closely at the amyloid plaques that collect during the course of brain aging and the development of Alzheimer's disease, we can see the central core area consists of insoluble amyloid protein. Around the outer rim are traces of inflammation. The theory is that our brains mount an inflammatory response attack against the amyloid protein, attempting to get rid of it, and this inflammation causes cell death and memory loss.

Scientists are focusing their efforts on developing specific drugs to target the brain's inflammatory response. Dr. Michael Mullan and his colleagues at the University of South Florida recently described how microglia cells promote an immune response that causes inflammation of the brain by releasing proteins known as cytokines, which are toxic to brain cells. The researchers found that the receptors involved in this response were present in the microglia cells of the brain's frontal lobe and the hippocampus—the areas affected by Alzheimer's disease. Although these receptors can also be found in the brains of people who are simply aging normally, the discovery still adds weight to the inflammatory hypothesis. If the current studies prove successful, people could choose to take precautionary anti-inflammatory drugs for decades before reaching an age when Alzheimer's is likely to strike.

Recently, Dr. Anthony Broe and colleagues at the University of Sydney, Australia, found that low doses of anti-inflammatory drugs worked just as well as higher doses for lowering the risk for Alzheimer's disease. Of the nearly 140 aspirin users in their study, 80 percent took only a half tablet of aspirin each day. This low dose would be insufficient to mount an anti-inflammatory response in the brain, which raises the possibility that interrupting the inflammatory reaction might not be the critical mechanism in protecting the brain.

Blowing the Smoke Out of Nicotine: Transdermal Patches

Most of us know about the negative consequences of smoking, particularly lung cancer and heart disease. However, due to the damage it does to the brain's circulation, smoking can also increase the risk for age-related memory loss.

Ironically, nicotine receptors in the brain are important for memory performance because they respond to neurotransmitters such as acetylcholine (tiny brain messengers carrying information we wish to remember). In Alzheimer's disease, these same nicotine receptors decline and die out, making learning and recall more difficult.

Recent studies have shown that if nicotine can be delivered to the brain through the skin using transdermal patches or another method, thus avoiding lung, mouth, and throat exposure, short-term memory performance improves, especially in people with only mild memory losses. The diminished nicotine receptors in Alzheimer's disease patients may limit the usefulness of nicotine patches because these patients may have too few receptors left to augment. The long-term memory benefits of this approach are currently being studied.

Plaque Busters and Detanglers: Vaccines and Other Treatments That Target Amyloid

The growing knowledge of what constitutes the basic brain lesions of Alzheimer's disease, particularly amyloid plaques and tangles, has thankfully propelled drug research and discovery toward prevention. The "plaque busters" that could one day reverse the deviant protein structures forming in our brains include synthetic proteins to break up the sheets of amassed amyloid, and vaccines consisting of synthetic amyloid-beta, the building blocks of the insoluble amyloid proteins that lie at the heart of amyloid plaques. Several pharmaceutical companies are currently developing molecules that inhibit production of the enzymes that lead to the buildup of the amyloid-beta protein building blocks, and human testing is just now beginning.

Perhaps the most promising breakthrough of late is a vaccine developed to create a heightened immune response that sloughs off the plaque deposits leading to Alzheimer's disease. The immune system, or the body's mechanism for fighting off disease, recognizes antigens or foreign bodies and responds with antibodies designed to search and destroy the foreigner.

Dr. Dale Schenk and his associates at Elan Pharmaceuticals have been working on a system that may stop the plaques from forming in the first place by attempting to jog the brain's immune system memory. The researchers vaccinated mice with a synthetic form of amyloid-beta that had been genetically engineered to form Alzheimer's plaques. Conventional vaccinations are designed to give the body's immune cells a taste of a particular infection that may come at it in the future. Afterward, should the real infection develop, these cells are already primed to produce antibodies to that infection more quickly than they would have otherwise.

In Dr. Schenk's study, the scientists found that monthly injections of this protein raised the level of antibodies the brain produced and actually prevented Alzheimer's disease–type brain degeneration in young Alzheimer's transgenic mice—ones that were genetically programmed to produce the amyloid-beta that creates the sticky plaques of the disease. Remarkably, the injections also eliminated plaques in older mice by as much as 80 percent. The vaccine improved their cognitive ability to master a maze.

Although the vaccine already appears safe in its initial human studies, experts remain uncertain as to whether the approach will stave off Alzheimer's disease. The Alzheimer's mice differ from humans with Alzheimer's disease. These animals do not show many of the symptoms seen in humans suffering from the disease, including the death of nerve cells. We don't yet know if the memory loss associated with Alzheimer's disease will significantly improve following depletion of the amyloid plaques, and it is possible that immunization with amyloid-beta may simply clear the plaques from the brain and have only minimal effects on memory loss.

The only sure-fire way to know if the vaccine effectively treats the disease is to test it in humans in a large-scale, double-blind placebo test. If the vaccine does work, it will be revolutionary. Not only might it prevent the disease in healthy people at risk for Alzheimer's disease but it may also lessen the existing symptoms in patients already suffering from the disease. Early candidates for determining if the vaccine prevents Alzheimer's disease would likely be people with a family history of Alzheimer's disease, the APOE-4 genetic risk, or both.

Studies of the vaccine in Alzheimer's patients are currently under way. Patients are receiving four injections over a six-month period to assess safety and measure buildup of the antibodies that will hopefully rid the brain of plaques. The hospitals where the initial studies took place were kept secret so the public would not besiege the investigators with requests to sign up for the trials.

Another recent report on a possible plaque buster treatment for Alzheimer's disease involves an older, lesser-known antibiotic called *clioquinoline*, once used to treat traveler's diarrhea. Dr. Ashley Bush and colleagues at Harvard Medical School screened dozens of antibiotics and anti-inflammatory drugs to find one that attached to copper and zinc, prominent components of the Alzheimer's amyloid plaques, in order to develop a treatment that would eliminate these metals from the plaques. When Alzheimer's transgenic mice received clioquinoline, the antibiotic attached itself to the metals in the brain plaques and cleared them out, leading to more than a 50 percent reduction in plaques. The mice also improved in their general behavior. Clioquinoline testing in human Alzheimer's disease patients is currently under way.

Making Headway in Healing: Growing Brain Cells, Gene Therapy, What Lies Ahead

The not-too-distant future may bring interventions for memory loss that today would seem like science fiction. Although neuroscientists have long believed that humans cannot generate new brain cells, recent findings now contradict this belief. Dr. Fred Gage of the Salk Institute in La Jolla, California, has done a series of studies showing that new nerve growth, or neurogenesis, is possible in adult human brains.

Gage's team established methods by which to isolate dividing progenitor cells from the adult brain and examine them in laboratory dishes. These progenitor cells are primitive, undifferentiated cells that can develop into specialized ones. The scientists' ability to extract and propagate these cells has led to wide acceptance of the possibility of neurogenesis in the adult brain. The next step is to grow these cells and use them as replacements in brains with diseased cells, like in Alzheimer's-affected brains.

Several scientific groups are developing stem cells, unspecialized cells that eventually turn into all specialized tissues in the body. Dr. Daniel Geschwind and Dr. Harley Kornblum of UCLA recently reported that neural stem cells can develop into any type of nervous system cell as well as into non-neural tissues. Their studies are building the technology that may one day make stem cells a viable approach to treating Alzheimer's disease and other brain disorders.

Fetal cell implants have been developed for growing new nerve cells in the brain; however, this approach had a recent setback in studies of Parkinson's disease. A carefully controlled study attempting to treat Parkinson's disease by implanting cells from aborted fetuses into patients' brains gave relief to some patients by partially improving their rigidity and slowed movements. Unfortunately, approximately 15 percent of the patients had too much cell growth, and a year or so after the surgery they produced an excess of the chemical that controls the movement problems in the disease. These patients ended up experiencing uncontrollable writhing and jerking.

The control group in this study had "sham" or fake surgery to make sure that the general aspects of the surgical procedure did not mask any potential benefit from new brain cell growth. Although critics argue for halting further studies using this technology, advocates note that successfully demonstrating fetal cell growth in a patient's brain is a major step forward. These investigators believe that with further technical refinement neurogenesis may eventually prove to be an effective approach to treating Parkinson's and Alzheimer's.

With genetic decoding, cloning, mapping and other discoveries in the last decade, investigators are working to apply these recent genetic insights to practical interventions in patients with Alzheimer's disease or people at risk. Scientists at the University of California, San Diego, recently made some headway in this direction when they performed the first surgery using gene therapy treatment on an Alzheimer's patient. The rationale for beginning gene therapy studies in humans is based on previous work in laboratory animals showing that nerve growth factor gene therapy prevents the death of cholinergic cells and reverses cell aging.

Although not intended as a cure, the therapy may protect or even restore some brain cells and relieve symptoms such as short-term memory loss. In the procedure, Dr. Mark Tuszynski and his co-workers took skin cells from the patient and genetically engineered them in the laboratory to produce and secrete human growth factor. The genetically modified cells were then surgically implanted in the brain's frontal lobe, an area involved in cholinergic neural transmission, memory processing, and reasoning. The scientists intend to determine whether preventing degeneration of this cholinergic system can improve memory performance in Alzheimer's patients and whether the nerve growth factor could prevent cell death.

The hope is that the implants will take hold and stabilize in humans; however, there are potential risks. For instance, the implants could begin dividing like tumor cells, thus raising safety issues about the procedure in general. One of the challenges in treating Alzheimer's disease has been to find drugs that can cross the blood-brain barrier and reach the area of the brain that is affected by the disease. If this gene therapy procedure proves to be successful, it will open the possibility of delivering drugs directly into the brain.

Other new drugs in the pipeline aim to boost the brain's ability to form memories despite the plaques. A group of drugs called *ampakines* increase the activity of brain chemicals important to memory formation. Ampakines are being tested in patients with Alzheimer's disease as well as people with only mild cognitive impairment. Additional drugs stimulate the production and release of growth factors in the brain's memory centers. Such growth factors coax nerve cells to create new connections with each other. Some pharmaceutical companies are developing experimental drugs that enhance the activity and possible growth of undamaged neurons in the brain's memory centers. Even if a successful approach is developed to eliminate plaque buildup, patients will still need to have treatment of memory problems. Thus, one medicine or course of treatment may affect disease progression while another improves function.

As newer technologies emerge in the next decade, I predict even more profound breakthroughs to halt the devastating march of brain aging. I am convinced that our current technological capabilities will lead to major breakthroughs within the next five to ten years. In the meantime, available drugs for memory, depression, and physical illness do have an important impact on brain health and offer the potential for staving off further memory decline.

Using Medicines Wisely for Keeping Your Brain Young

- **Learn how doctors organize information so you can be proactive in your medical evaluation.**
- **Numerous physical illnesses can further age your brain. Take them seriously—see your doctor sooner rather than later.**
- **Avoid using more medicines than you need. Talk with your doctor about any drugs you are taking that could influence memory ability.**
- **Remain cautious about taking herbal supplements.**
- **Hold off on using new or innovative treatments until results from clinical trials are in.**
- **Remember that treating a depression with the right antidepressant drug often improves memory performance.**

Chapter Ten

Don't Forget the First Nine Chapters

Why put off until tomorrow what you can forget to do today?
—Gigi Vorgan

Some cynics, my wife included, may ask: "If we're already suffering from brain aging and annoying daily forgetfulness, how are we ever going to remember the last chapter, or any of the strategies for keeping our brains young?" Well, to you, dear, and others of your ilk, I submit this chapter as a practical guide to pull together your entire memory program and keep you on track as you progress.

Pick and Choose

No matter how well we learn mnemonic techniques, there will always be too much information to remember. Even people with savant skills for memorizing lists of trivia have limitations in their ability to store facts and figures. Many of the people who succeed in learning and recall skills, and life in general, have learned to *choose* which information is useful to learn and which information is less important and can be glossed over.

This selective process requires a conscious effort at first but can become second nature with practice, and its usefulness is immeasurable. It might be a good idea to remember your boss's birth date, but the date of his hip replacement is not one you necessarily need to commit to your long-term memory stores.

Once you have selected information you know you want to remember, then you can choose the best memory tool, whether it's *LOOK, SNAP, CONNECT* (Chapter 3), the Peg Method (Chapter 6), or other tools at your disposal. Sometimes you can choose not to necessarily remember a fact outright, but to write it down, enter it into your date book, put it on your to-do list, or pass the buck and ask someone else to remember it.

Be sure to set achievable goals when picking and choosing which information to remember outright, and which information requires use of an *internal* memory tool (e.g., mnemonic technique) or *external* memory tool (e.g., reminder note). I know from embarrassing experience that I need to consult my son's school roster before attending his first-grade open house because I do not know the names of his classmates' parents by heart, even as the end of the semester approaches. My wife does, and I'm sure she's gloating about it as she reads this.

Getting Organized to Expand Your Memory Power

Beginning a memory program that encompasses as many methods and tools as we have learned may seem overwhelming unless we organize them into an easily implemented system. The following are some organizational approaches to help keep your memory abilities at peak performance.

- ***Write effective notes.*** I first learned about summary notes in high school. My senior English teacher had us look over each page of notes and write a brief summary at the top. The process of summarizing the information forced us to think about what we had learned that day, condense it, and rewrite it in different words. This process helped fix the information into my memory. Well-written summary notes condense the amount of material we have to remember, and sometimes the simple act of writing things down helps facilitate recall. The more thought and effort we put into creating a note of something to remember, the more helpful it will be.

- ***Organize memory places.*** One of the most common memory complaints is forgetting where we put things. An effective option for avoiding the "disappearing keys act" is to put commonly misplaced items in the same "memory" place—a hook near the door in the kitchen for your car keys, the same briefcase pocket for your organizer, and that convenient desk drawer for the scissors and the pencils.

Your office, home, and car can be more efficient if the storage areas and living spaces are organized with designated memory places. When my family moved across town several years ago, I was struck by how many things I could still not find in the new house, even after months. A major problem was that I had lost all my old memory places for various items, from light bulbs to tools, and I had not adapted an organized memory place system in the new house. This kind of strategy involves planning. For example, you might keep writing utensils in a convenient place in the study or den but ski equipment and other seasonal items stored out of the way. Some of us do this instinctively, while others take the concept of the "junk drawer" to its furthest household-encompassing limit and might benefit by creating "memory drawers."

- ***Use daily planning lists.*** I learned about this technique while I was a busy first-year resident in internal medicine. Each of us had to care for a dozen or more hospitalized patients who were acutely ill. The only way to keep track of all the patient-care tasks was to make lists. I have continued to use this memory tool ever since. If you don't already use daily planning lists of things to do, I urge you to try them. Once you complete a task, cross it off. After a few days, transfer the active items to a new list. When the lists get long or complex, place asterisks next to items that need more immediate attention.
- ***Use weekly or monthly planning calendars.*** A wall or desk calendar is a helpful way to keep track of regular or occasional events or meetings. Our family has one displayed prominently in the kitchen, and we have gotten in the habit of making sure important appointments and weekly activities are posted.
- ***Get a date book or try an electronic hand-held organizer.*** Pocket date books have long helped busy people keep track of the details in their lives. For many people, the newer electronic pocket gadgets have replaced date books because of their many programs, including calendars, phone books, and to-do lists, as well as Internet access. You can download the information onto your desktop computer so that, should you lose your hand-held organizer, you won't lose all the information you've entered into it. You can also print out your schedule, phone book, and lists for others.
- ***Use Post-its for quickies.*** Many people prefer Post-its or stick-on reminder notes as external memory aids. If you haven't tried them, you might consider using them to augment your other strategies. They are a good quick fix

as a reminder but try not to overuse them. An inherent risk in using stick-on notes is that sometimes they "unstick." Either follow up on the thought or task or transfer it to a more stable memory tool.

- ***Develop memory habits.*** From the time we are children, we learn memory habits. We brush our teeth morning and night. We take vitamins after breakfast (I do, especially vitamin E, 400 IUs—see Chapter 7). When the dentist notes your chronic incapacity to floss, you might create a new memory habit by placing the floss next to the toothpaste. I've noticed that if I ask my wife to drop my shirts at the cleaners, she scowls, then routinely puts them in the front passenger seat of her car. This is her memory habit to remind herself to get to the cleaners. Daily pill box organizers, alarm clocks, watches, and other tools are available to augment memory habits.
- ***Plan a daily routine.*** We all do better with a certain amount of structure in our lives. Many of us have had the experience of waking up in a hotel room and momentarily experiencing confusion about where we are. It usually takes a few seconds to recall that we're not at home but in a hotel. If a person has Alzheimer's disease or another serious memory problem, changing their daily activities randomly can make them more confused. If we build a general routine into our daily schedule, we will have more time to focus on work, leisure, and others things we want to learn.
- ***Don't overdo it with memory aids.*** Too many lists, Post-its all over your dashboard, and a date book so jam-packed that you can't make out who you are lunching with won't necessarily be useful or effective. Taking random, copious notes that are never read tend to be a waste of energy.

On the other hand, succinct lists, notes, and reminders can be extremely helpful memory tools, and I sometimes keep mine for years. Picking and choosing the memory tool can be as important as picking and choosing the information you wish to remember.

An Innovative Strategy for Keeping Your Brain Young: Review and Organize

Jot down the ten chapters of *The Memory Bible* and post them prominently (see box).

> **An Innovative Strategy for Keeping Your Brain Young**
>
> **Chapter 1: You Have More Control Than You Think**
> **Chapter 2: Rate Your Current Memory**
> **Chapter 3: *LOOK, SNAP, CONNECT:* The Three Basic Memory Training Skills**
> **Chapter 4: Minimize Stress**
> **Chapter 5: Get Fit with Mental Aerobics**
> **Chapter 6: Build Your Memory Skills Beyond the Basics**
> **Chapter 7: Start Your Healthy Brain Diet Now**
> **Chapter 8: Choose a Lifestyle That Protects Your Brain**
> **Chapter 9: Wise Up About Medicines**
> **Chapter 10: Don't Forget the First Nine Chapters**

Staving off brain aging requires consistency and long-term commitment, but no one single memory program meets everyone's needs. Each of us has areas of strength and weakness from the outset. If you're already an athlete or jog four days a week, but you can't seem to break your penchant for cheeseburgers, French fries, and beer, then you should focus more attention on adopting a healthy brain diet rather than a brand-new physical fitness program. If you have a stressful job, family life, or both, then minimizing stress and anxiety may take precedence over designing your optimum mental aerobics program. You're probably getting plenty of mental aerobics just keeping up with your job. Many people are able to work on several different strategies in their memory improvement program if they can organize the individual tasks in a way that fits into their daily routine.

To help you tailor your memory improvement program to your particular needs, the following is a brief summary of the first nine chapters. As you review them, check off key points in the shaded boxes that you feel you may want to focus extra attention upon, be it mental aerobics, memory training, diet, or physical exercise.

Chapter 1: You Have More Control Than You Think

We have all come to expect and accept occasionally forgetting our keys or people's names, but new and compelling evidence shows that early age-related forgetfulness is actually the brain's first warning sign of its gradual decline. Recent scientific discoveries show not only that we can begin to detect subtle early evidence of brain aging but that we can do something about it. By combining brain scans with recently discovered genetic markers, we can pinpoint the earliest indications of impending brain aging. We also now know that our brains have the power to fight back with an easy yet comprehensive program of memory training and brain fitness. If we accept that brain aging is a lifelong process, then why not embrace a memory fitness program to keep our brains healthy as a lifelong commitment? It's never too late or too early to protect our brain cells and delay memory decline.

Chapter 2:
Rate Your Current Memory

To begin a memory-training program and set reasonable goals, we need to rate our current subjective and objective memory abilities. Remember that *subjective* memory is our own perception of how well we think we do in memory functions, while *objective* memory is how well we actually perform on a pencil-and-paper memory test.

Go back to the *subjective* memory assessment in Chapter 2 and complete the questionnaire, using a different-colored pencil. Compare your current score with your earlier one (fill in chart in Figure 2.1), and you may be pleasantly surprised at the progress you've made already.

Afterward, reassess your *objective* memory by studying and recalling the new list of words in Assessment No. 3. Check your watch or timer before starting.

ASSESSMENT NO. 3
STUDY THE FOLLOWING WORDS FOR UP TO 1 MINUTE:

Arrow
Pepper
Elephant
Stain
Toast
Instructor
Cigar
Grandmother
Hammer
Swamp

Now put aside *The Memory Bible* and reset your timer for a 20-minute break. Do something else—water the plants, check your e-mail, whatever you like. After 20 minutes, write down as many of the words as you can recall. Compare your score with your earlier ones recorded in the chart in Figure 2.1.

As you continue with your memory program beyond the reading of this book, you can refer back to these ratings and follow your improvement. If you stick with your program over time, you should continue to see results. At the very least, you should be able to maintain your earlier achievements and keep one step ahead of brain aging.

Chapter 3: *LOOK, SNAP, CONNECT*: The Three Basic Memory Training Skills

My approach to memory training comes down to three basic skills—*LOOK, SNAP, CONNECT*. That's it. *LOOK, SNAP, CONNECT.* If you learn these three basic skills, then I probably won't be seeing you at my memory clinic any time soon.

- ☐ **LOOK—*Actively Observe What You Want to Learn.* Slow down, take notice, and focus on what you want to remember. Consciously absorb details and meaning from a new face, event, or conversation.**
- ☐ **SNAP—*Create Mental Snapshots of Memories.* Create a mental snapshot of the visual information you wish to remember. Add details to give the snaps personal meaning and make them easier to learn and recall later.**
- ☐ **CONNECT—*Link Your Mental Snapshots Together.* Associate the images-to-be-remembered in a chain, starting with the first image, which is associated with**

the second, the second with the third, and so forth. Be sure the first image helps you recall the reason for remembering the chain.

Chapter 4: Minimize Stress

Managing and reducing the chronic stress so many of us experience in our busy, frenetic lives is likely to slow brain aging and improve our physical health. Relaxation exercises and avoiding unnecessary outside stressors can lead to diminished anxiety and better recall.

Mastering internally driven stress may involve altering the way we behave, the way we think, or the lifestyle choices we make (Chapter 8). Chapter 4 describes several ways to minimize daily stress and anxiety:

- ☐ Cut back on caffeine.
- ☐ Exercise regularly.
- ☐ Prepare ahead to avoid stress.
- ☐ Learn how to relax and do it at regular intervals.
- ☐ Get enough sleep.
- ☐ Pace your day.
- ☐ Balance work and leisure.
- ☐ Set realistic expectations.
- ☐ Talk about feelings.
- ☐ Let yourself laugh.
- ☐ Get treatment for anxiety and depression if necessary.

Chapter 5: Get Fit with Mental Aerobics

Research points to mental stimulation and brain training as a way to keep our brains young and healthy. Practicing mental aerobics using a cross-training approach, which varies our brain-training routine day by day, minimizes boredom and maximizes results.

It's important to start your mental aerobic exercises at the correct level of difficulty for you. Mental stimulation exercises should be challenging and enjoyable to achieve their best effect.

Some people are naturally more inclined toward right-brain skills (e.g., spatial relationships, artistic and musical abilities, face recognition, depth perception) and may need extra work on their left-brain skills (e.g., logical analysis, language, reading, mathematics, symbol recognition). By contrast, others have the opposite balance of abilities.

Use your score on the subjective and objective memory assessments (Figure 2.1) as a guide to determine where best to begin to focus your mental aerobics program:

- ☐ **Beginning exercises (low memory performance scores)**
- ☐ **Intermediate exercises (middle memory performance scores)**
- ☐ **Advanced exercises (high memory performance scores)**
- ☐ **Left-brain exercises**
- ☐ **Right-brain exercises**

You can find additional mental aerobics exercises through the Internet or in your local bookstore.

Chapter 6:
Build Your Memory Skills Beyond the Basics

Although *LOOK, SNAP, CONNECT* provides quick memory-improvement results, many of us would like to take these skills further. Review some of the more advanced memory-training skills below that you would like to practice and use:

- ☐ *Organization.* Look for systematic patterns and groupings to facilitate learning and recall.
- ☐ *Peg Method for Remembering Numerical Sequences.* Commit to memory a specific visual "peg" for each of the ten digits and then use the link method to create a story to remember any numerical sequence.
- ☐ *Remembering Names and Faces.* Make sure you consciously listen and observe the name (*LOOK*), then *SNAP* and *CONNECT* to fix the name to the face.
- ☐ *Roman Room Method.* Pick a familiar room or route and in your mind place the items to remember at key points or landmarks.

Chapter 7: Start Your Healthy Brain Diet Now

The sooner we start our healthy brain diets, the sooner we will reap the benefits. Note each of the dietary adjustments below that you would like to make:

- ☐ Drink at least six glasses of water a day.
- ☐ Plan your meals in advance, try to keep your portions low, and eat healthy between-meal snacks.
- ☐ Brush your teeth a few hours before bedtime as a reminder to avoid evening snacking.
- ☐ Get unhealthy stress foods out of your house, car, and office. Try substituting with snack bags of fresh cut vegetables when you need to munch.
- ☐ If you catch yourself in a stress-eating mode, hit your pause button. Take only one bite of that cookie, candy bar, or crème brûlée, and take a deep breath. Try stretching for a few minutes.
- ☐ Eat a low-fat diet that includes plenty of fruits and vegetables.
- ☐ Avoid processed foods and high-glycemic-index carbs.
- ☐ Eat foods rich in omega-3 fats and avoid omega-6 fats.
- ☐ Avoid too much caffeine.
- ☐ Take a multivitamin, vitamin E, vitamin C, and folic acid supplements.

Chapter 8: Choose a Lifestyle That Protects Your Brain

Whether or not we age successfully stems largely from our daily lifestyle choices and the environment in which we live them. In fact, lifestyle and environmental factors outweigh genetic factors by a 2-to-1 ratio. Most of us know it's best to avoid smoking, sleep deprivation, and head trauma to maintain physical and mental health, but many of us don't realize how many other daily lifestyle choices have long-lasting effects. Note any of the lifestyle choices below that you would like to incorporate, and start your own list of positive lifestyle changes you can make to begin protecting yourself from Alzheimer's disease.

- ☐ Start an exercise program to maintain aerobic fitness and flexibility.
- ☐ Get the aerobic and social benefits of walking with friends several times a week.
- ☐ Choose sports and physical activities with low risk for head trauma and always wear a helmet when riding a bike.
- ☐ Don't drink and drive and always wear your seat belt.
- ☐ If you smoke, quit. Ask your doctor if you need help.
- ☐ If you drink alcohol, do it in moderation.
- ☐ Get out and stay involved in activities that have personal meaning. Spend time with friends and family.
- ☐ Get plenty of sleep.

Chapter 9: Wise Up About Medicines

As new scientific technologies and pharmaceutical discoveries continue to emerge in the next decade, we can expect even more profound breakthroughs in our fight against brain aging and Alzheimer's disease. However, right now, we have drugs available for memory, depression, and physical illness that have an important impact on memory performance and brain aging. To use these medicines wisely, it is important to keep in mind the following:

- ☐ **Learn how your doctor organizes information about your health so you can become more proactive in your medical care.**
- ☐ **Physical illnesses can threaten brain fitness. Take them seriously—see your doctor sooner rather than later.**
- ☐ **Avoid using too many medicines if you can. Ask your doctor about any medicine you feel is influencing your memory ability.**
- ☐ **Remain cautious about taking herbal supplements.**
- ☐ **Hold off on trying new or innovative treatments until conclusive clinical trial results are in.**
- ☐ **Treating a true depression with the right antidepressant drug often improves memory impairment.**

Chapter 10: Don't Forget the First Nine Chapters

Keeping our brains young and protected against Alzheimer's disease involves attention to nearly all the areas of our lives. As a practicing psychiatrist, I know firsthand how difficult it is for people to change their habits. As a father, husband, and son, I also know of the challenges of helping others we care about make positive changes, as well as making changes for ourselves. By reading this book you have begun your commitment to protect your brain, but your personal motivation will be the driving force behind your memory program for keeping your brain young.

To give you an idea of how to put a memory program together, here is an example of one person's initial schedule.

How It's Done: Putting *The Memory Bible* to Work

A senior executive in a manufacturing company had the flexibility in his schedule to keep up with several aspects of a brain fitness program, including memory training, mental aerobics, and physical exercise. He also worked hard during his first week to cut back his coffee consumption, although he did need to take Tylenol for several days to help with his caffeine-withdrawal headaches.

Because he felt he got enough mental aerobics on the job and through his daily reading, he used his time to work on memory-training techniques rather than specific mental aerobics exercises. For him, minimizing stress and anxiety was a greater challenge. Here's how his memory program shaped up the first few days.

Sample Memory Program

Day 1

CHAPTER	ACTIVITY
3-Memory training	Active observation memory-training exercises (10 minutes)
4-Minimize stress	10-minute yoga session at bedtime 20-minute afternoon work break
5-Mental aerobics	Read *New York Times*, did crossword at breakfast
7-Diet	Multivitamin (containing 400 micrograms folic acid), vitamin E (800 IU), vitamin C (1,000 mg)
	6 glasses of water
	Breakfast—sourdough toast, non-fat cheese slice, egg white, 2 cups of coffee
	Lunch—salad with low-cal dressing, ham, 1 cup of coffee
	Dinner—salmon, steamed broccoli, non-fat yogurt dessert
	Snacks—string cheese and blueberries
8-Physical exercise	Stairs 3 times at work
	Walked dog briskly after dinner—6 blocks
9-Medicine	Anti-hypertensive, antidepressant

Day 2

CHAPTER	ACTIVITY
3-Memory training	Mental snapshot memory-training exercises (10 minutes)
4-Minimize stress	10-minute meditation session during lunch break
	Early to bed (9:30 P.M.)
5-Mental aerobics	Read *New York Times*, did crossword puzzle
7-Diet	Multivitamin, vitamin E (800 IU), vitamin C (1,000 mg)
	7 glasses of water
	Breakfast—sourdough toast, non-fat cheese slice, egg white, 1 cup of coffee
	Lunch—Non-fat yogurt, banana, tuna, orange, 1 cup of coffee
	Dinner—skinless chicken, rice, steamed vegetables, 1 glass of red wine, apple pie
	Snacks—cut vegetables, cottage cheese, ½ cup of coffee
8-Physical exercise	Stairs 2 times at work
	Paddle tennis at the sports club
9-Medicine	Anti-hypertensive, antidepressant, Tylenol (2)

Day 3

CHAPTER	ACTIVITY
3-Memory training	Association and linking memory-training exercises (10 minutes)
4-Minimize stress	10 minutes meditating after discussing complicated personnel issue with co-workers
5-Mental aerobics	Math puzzle book; played word game with daughter
7-Diet	Multivitamin, vitamin E (800 IU), vitamin C (1,000 mg)
	7 glasses of water
	Breakfast—Bran cereal, non-fat milk, grapefruit, 1 cup of coffee
	Lunch—Tuna sandwich, tea, strawberries
	Dinner—skirt steak, potato, steamed vegetables, 1 glass of red wine, non-fat yogurt
	Snacks—blueberries, string cheese, ½ cup of coffee
8-Physical exercise	Stairs 4 times at work
9-Medicine	Anti-hypertensive, antidepressant, Tylenol (4)

Starting *Your* Memory Improvement Program

As in the example above, you may wish to follow a few, several, or all of the strategies at the outset of your memory program. You may want to incorporate the items you checked off in the shaded areas above in designing it. Some people may simply want to begin with a to-do list of tasks, such as:

- **Get junk food out of the house.**
- **Buy small water bottles and distribute throughout house, car, office.**
- **Purchase Post-its.**
- **Talk to brother-in-law about getting a hand-held organizer.**
- **Focus on left-brain mental aerobic exercises like word jumbles and crosswords.**

The key is to be organized in adjusting your memory program to complement your schedule. Some of the organizational strategies described at the beginning of this chapter, such as planning lists and weekly calendars, can be particularly helpful.

On the following pages you'll find examples of how different people organize their memory programs. The person who contributed the worksheet was detailing his progress in both mental aerobics and stress reduction. The other individual used his computer's daily calendar to record both positive achievements as well as areas where he had slipped during the day. You can create your own worksheet or calendar records, possibly emphasizing areas where you feel you need extra focus. For example, you might keep track of your daily fat intake, hours of sleep, and enjoyable activities you engage in each week outside the house.

Mental Aerobics and Stress Reduction Worksheet

TASK	Monday	Tuesday	Wednesday	Thursday	Friday
Mental aerobics	10 minutes	5 minutes	skipped	10 minutes	5 minutes
Stress-reducing exercise	Treadmill 20 minutes	Tennis with daughter	Power walk	Stairmaster–30 minutes	Power walk
Relaxation	TV after dinner	Read before bed	Took a 2–hour lunch	Read before bed	Tried meditation
Caffeine	2 cups coffee	1 1/2 cups coffee	1 1/2 cups coffee	1 cup coffee	1 cup coffee
Sleep	5 1/2 hours	5 hours	7 hours	7 1/2 hours	8 hours
Work/leisure ratio (hrs.)	8/6	11/5	9/6	8/6	7/7

June 19

TUESDAY

JUNE						
S	M	T	W	T	F	S
					1	2
3	4	5	6	7	8	9
10	11	12	13	14	15	16
17	18	19	20	21	22	23
24	25	26	27	28	29	30

JULY						
S	M	T	W	T	F	S
1	2	3	4	5	6	7
8	9	10	11	12	13	14
15	16	17	18	19	20	21
22	23	24	25	26	27	28
29	30	31				

8:00	Bagel, coffee, cheese omelet for breakfast
9:00	
	Oops! Ate coffee cake with 2nd cup of coffee
10:00	2 glasses of water, took 4 flights of stairs
11:00	Work break—15 minutes of brain-teasers
12pm	Tuna salad, green tea, salad for lunch (repent for coffee cake)
1:00	Yelled at colleague (took 10-minute relaxation break to cool down)
2:00	
3:00	Power walk into village for aerobic workout
	Fruit and cheese snack
4:00	Meet with Jim about new hand-held organizer
5:00	1 glass of water
6:00	
7:00	Dinner, salmon, tomato soup, too much wine—3 glasses. Sherbet dessert only!!!
8:00	
	1 glass of water
9:00	Brush teeth early to avoid snacks
	Relax time—watch mindless TV

Keeping Your Brain Young for Life

You have read *The Memory Bible* and now have many tools to help keep your brain young, protect against future memory decline, and delay, or possibly even ward off Alzheimer's disease. The Appendixes that follow include very recent discoveries on detecting and treating brain aging and Alzheimer's disease, as well as useful information for family members and caregivers of those with dementia or Alzheimer's, including new and available medications and nonmedical strategies.

Because there are so many new potential treatments, including drugs, hormones, herbs, and even surgical treatments, I have included, in Appendix 3, a description of the more popular ones to help readers sift through what has been shown to be effective and what has not. The Appendixes also provide a glossary of terms and additional resources to help you build and continue building your program for the rest of your life. Good luck!

APPENDIX 1

The Amyloid Probe—Keeping Watch on Plaques and Tangles

As I was finishing up *The Memory Bible*, our UCLA research group discovered an innovative method to view the Alzheimer's amyloid plaques and tangles directly as they accrue in the brains of living people—thus avoiding the unnecessary delay, trauma, and, of course, death that is required by the conventional method of viewing these brain lesions only at autopsy. This new discovery not only provides investigators a way to monitor new drugs for treating and preventing Alzheimer's disease, but it could help us detect the disease earlier and get a jump on thwarting it altogether.

After combining APOE genetic and PET scan information to better detect and treat Alzheimer's disease and those people at risk for developing it, I brought together a few UCLA colleagues from diverse scientific backgrounds, in an effort to find an even more specific approach to imaging the Alzheimer's brain—a method of visualizing plaques and tangles.

Our small group made quick progress. Within a few meetings, Dr. Greg Cole, a neuroscientist with extensive experience in Alzheimer's disease, made clear that the internal environments of plaques and tangles were hydrophobic, that is, more friendly to fat than to water. Dr.

Jorge Barrio, a renowned chemist, had recently synthesized a new group of compounds that thrived in these hydrophobic environments, and these molecules passed easily from the blood stream to brain tissues.

We began our initial studies by using these new compounds on autopsied Alzheimer's brain tissues. These experiments were successful in clearly displaying the well-defined amyloid plaques and tangles characteristic of the disease. We then moved on to injecting the compounds into living Alzheimer's patients, followed by PET scans. This allowed us to see, or probe, for the first time, increased signals coming from living human brains in areas that contained dense collections of plaques and tangles. Seeing the lesions with this new *amyloid probe* gives scientists the ability to monitor plaque and tangle concentrations while testing experimental treatments to eliminate them.

The amyloid probe essentially seeks out and temporarily attaches itself to the plaques and tangles, thus providing a clear PET scan signal in the areas of the brain where Alzheimer's strikes first, the hippocampus and temporal cortex. In healthy people without Alzheimer's, these regions produce little or no signal. However, in people with the disease the signal is so strong and accurate that it actually correlates with each individual's degree of memory impairment.

Figure A.1 shows two different scans of a patient who suffers from Alzheimer's disease. The image on the left is an amyloid PET scan, while the image on the right is a conventional PET scan. Note that the temporal areas (arrows) are darker in the amyloid scan, indicating an increased signal—this is where the plaques and tangles accumulate. The conventional PET scan, which measures the functional ability of brain cells, indicates low activity in these same areas (lighter gray area).

Figure A.1

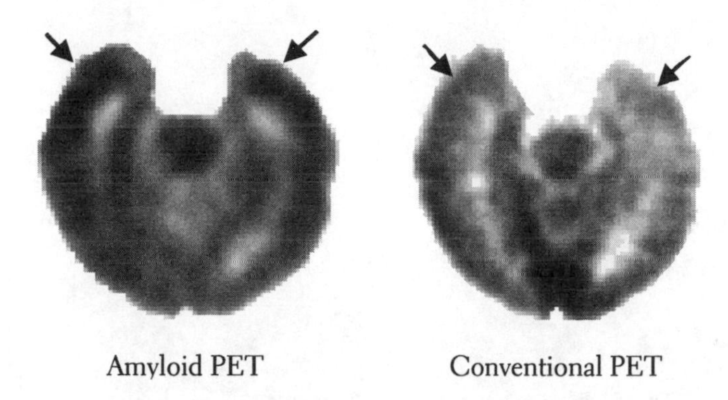

Our group is currently working with other researchers to use this amyloid probe technology to study medications being developed to wipe out plaques and tangles (Chapter 9). International enthusiasm for this technology is high since it promises to streamline drug discovery, particularly for treatments designed to slow and possibly eliminate age-related cognitive disorders and Alzheimer's disease.

APPENDIX 2

What to Do If Alzheimer's Disease Strikes

The Memory Bible will hopefully provide you an edge in the brain-aging game by introducing you to a mental and physical aerobics program, stress reduction techniques, and many other tools for maintaining brain fitness. If you bring the motivation, you can proactively stave off the advance of Alzheimer's as effectively and as long as possible.

However, in some cases the genetic risk is high or a person already has several existing risks, such as head injury, high cholesterol, or hypertension, and brain aging may have progressed to the point where it is interfering with daily life. Anyone in this situation should seek professional help.

Age is the single greatest-known risk factor for getting Alzheimer's disease. Approximately 5 percent of people age 65 years or older have the disease, but by age 85 that figure soars to between 35 and 47 percent. Advanced brain aging and Alzheimer's disease afflict 4 million people in the U.S. and nearly 25 million people worldwide. Despite such a high prevalence, Alzheimer's disease and other dementias remain under-recognized. Timely recognition is important because treatment is available. Both medication and non-medicinal interventions can slow the progression of the disease and improve functioning in most patients.

ALZHEIMER'S DISEASE: IT'S A FAMILY AFFAIR

The devastation of Alzheimer's disease doesn't stop with the patients, it spreads to their families and friends. Watching a loved one decline before your eyes, seeing the personality of someone you cherish gradually disappear, is a traumatic and confusing experience, often leading to anger, sadness, guilt, and depression in the family member and caregiver, as well as the patient. The physical characteristics of a person with Alzheimer's disease remain, but *who* they are eventually vanishes. I have had family members express relief when the patient finally dies, since they have been mourning the loss of that person little by little for years. Research has shown that more than 50 percent of caregivers develop depressions serious enough to require medical intervention. Caregivers miss days at work, have a high risk of becoming physically ill, and often lose sleep, especially when the patient's disease becomes more advanced and is accompanied by agitation and restlessness at night.

A diagnosis of dementia is a staggering consequence for patients and their families, many of whom may have a major economic burden to consider as well as this emotional blow. As baby boomers age over the next few decades, the number of older persons will rise steeply, as too will the cases of Alzheimer's disease. Sadly, by the year 2050 an estimated 14 million Americans will suffer from full-blown Alzheimer's disease. I predict those numbers could be much lower if more people followed at least some of the strategies in *The Memory Bible*.

Alzheimer's disease is the third most costly disease in the U.S., after cardiovascular disease and cancer. Annual costs exceed $100 billion, and most of that is not covered by medical insurance, leaving the families of Alzheimer's patients to bear the greatest economic burden. Earlier intervention in mild to moderate cases can enhance the daily functioning of patients and improve their quality of life. Clearly, keeping our brains young and avoiding symptoms of Alzheimer's disease should be the ultimate goal for all of us.

THE GRADUAL MARCH OF SYMPTOMS

Dementia is the general term doctors use to describe loss of memory and other cognitive functions when they impair daily life. With Alzheimer's disease, the course is gradually progressive, memory loss is usually the first symptom to appear, and motor and sensory functions are spared until late stages of the disease. Early on, patients have difficulty learning new information and retaining it for more than a few minutes. As the disease advances, the ability to learn is increasingly compromised, and patients have trouble accessing older, more distant memories. Patients develop problems finding words, using familiar tools and objects, and remaining oriented to time and place.

Eventually, all aspects of their lives become impaired: patients are unable to plan meals, manage finances or medications, use a telephone, and drive without getting lost, and these kinds of difficulties may be the patient's or family's first sign that something is amiss. Social skills usually remain until late in the disease, which contributes, of course, to the delay in recognition of the disease. As the dementia progresses, judgment becomes impaired, and patients have trouble carrying out even the most basic functions, like dressing, grooming, and bathing.

Sometimes families have to cope with personality changes, irritability, anxiety, or depression early in the disease, and as the patient worsens, these changes in mood and behavior become more common. Patients lose touch with reality and may become psychotic. They can experience delusions, hallucinations, and aggression, and often wander and get lost. These kinds of behaviors are the most troubling to caregivers, usually distress family members, and lead to nursing home placement.

Alzheimer's disease accounts for approximately 65 percent of dementia cases, while an estimated 15 percent have a condition of motor stiffness and rigidity similar to that found in Parkinson's disease. This combined dementia/Parkinson's syndrome has been termed "dementia associated with Lewy bodies," named for the small round abnormal accumulations found in the patients' brains. Patients with Lewy body dementia often have visual hallucinations and altered alertness. Still another form of dementia strikes primarily the front of the brain and the areas under the temples, so it is known as frontotemporal dementia. These patients often show marked changes in personality, have particular difficulty in executive skills and planning complex tasks, yet their visual and spatial memory tends to be preserved. The cumulative effect of multiple small strokes in the brain causes vascular dementia. Approximately 20 percent of patients with Alzheimer's disease also have vascular disease in the brain.

Many physical illnesses can cause dementia, including infections, cancers that spread to the brain, thyroid disease, or hypoglycemia. Chronic alcohol or drug abuse, as well as a variety of medicines, ranging from antidepressants to antihypertensives, over-the-counter drugs, sleeping pills and antihistamines, can also cause symptoms of dementia (Chapter 9).

MEDICAL SCREENING: HOW THE DOCTOR CHECKS YOU OUT

When those middle-aged pauses are no longer a joke and become a serious matter, *that* is when Alzheimer's disease can be diagnosed and hopefully recognized early. Too often, physicians and family members accept memory loss, symptoms of depression, and other important diagnostic clues as normal consequences of aging. In the early stages of dementia, I have heard about patients getting into a variety of complications: deeding their house away, being influenced about their will, losing track of substantial sums of money, or marrying a gold-digger.

A physician's evaluation of memory loss usually involves an interview, physical examination, and laboratory assessment. In evaluating the patient's mental state, the doctor will screen for depression, memory loss, and other cognitive skills. Laboratory assessments should at least include some blood tests to screen out thyroid disease, vitamin B_{12} deficiency, and other disorders, which could possibly cause memory change.

Doctors often obtain a standardized score of cognitive ability using rating scales like the Mini–Mental State Examination, which consists of 30 items that rate memory, orientation, attention, calculation, language, and visual skills. The test takes only about 10 minutes but is limited because it will not detect subtle memory losses, particularly in college graduates. More detailed memory assessments, known as neuropsychological tests, will provide a better idea about subtle memory deficits.

What's really important to families is how the patient is getting along at home. Dr. Ken Rockwood and his colleagues at Dalhousie University in Nova Scotia have developed a way to measure the patient's response to treatment in terms of how family members and caregivers assess them, focusing on those daily activities they find most important. The doctors created an individualized scale for each patient based upon the family's own descriptions. They ask what drives family members crazy about a patient, then create an outcome measure based on those reports. For example, if Susie T. complained that it drove her crazy when her father asked the same question twenty times a day, Dr. Rockwood might set the six-month treatment goal for Susie's father to asking the same question *only five times* each day.

BRAIN SCANS FOR EVALUATION OF DEMENTIA

Some experts recommend computed tomography (CT) or magnetic resonance imaging (MRI), which provide information on the brain's structure. These kinds of scans will detect strokes, brain tumors, or cerebral hemorrhages, which occasionally cause symptoms resembling dementia. Unfortunately, in most cases, MRI and CT provide only non-specific information about brain shrinkage or atrophy or white-matter changes, which show up as spots in the deeper brain areas. These changes are often difficult to interpret and rarely provide a diagnosis that will alter treatment.

By contrast, the positron emission tomography (PET) scan (Chapter 1) is the most effective way to arrive at an early diagnosis. PET scans enable a physician to make a positive diagnosis of Alzheimer's disease and of other types of dementia, facilitating the initiation of anti-dementia drugs, which can improve symptoms and slow the progression of disease while a patient still retains a high level of cognitive function. Early detection also gives patients and their families more time to plan for the future.

At a cost of approximately $1,200 per scan, PET can minimize the need for repetitive diagnostic tests. Depending on the particular clinical setting, such tests can cost upward of $2,000. In the long run, PET scans may save money because earlier diagnosis can eliminate lengthy, costly, and inconclusive evaluations.

Positron emission tomography can identify the Alzheimer's brain pattern months, even years before obvious symptoms of the disease appear in patients. Dr. Dan Silverman and others in our UCLA research group recently conducted an international study of nearly 300 patients focusing on the use of PET in the evaluation of dementia. We found that PET scanning is extremely sensitive to early changes in the brain and extraordinarily accurate in predicting the future course of dementia. It demonstrated nearly 95 percent accuracy in predicting the patient's clinical course over three years.

The images in Figure A.2 show two-dimensional brain slices with the front of the brain toward the top and the back of the brain toward the bottom. The arrows point to lighter gray areas of decreased brain function.

Figure A.2

PET SCANS SHOWING FUNCTION PATTERNS IN A NORMAL ADULT, AN ALZHEIMER'S PATIENT, AND A NEWBORN

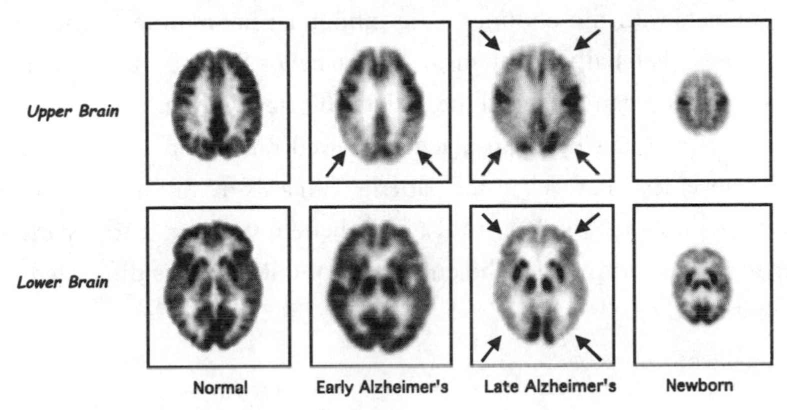

The scans reveal a consistent pattern in Alzheimer's disease. Those parietal (see arrows in early Alzheimer's) and temporal areas—where Alzheimer's first strikes—show reduced activity in the early stage of the disease. At the late stage—when patients have extreme trouble talking and interacting with others—the frontal areas show decreases. The dark areas midway between the front and back of the advanced Alzheimer's brain control sensation and physical movement and are still at work late in Alzheimer's disease, so these patients are able to experience sensation and control muscle movement. It is remarkable to note that the PET scan of a late-stage Alzheimer's patient looks very similar to that of a newborn.

SLOWING DECLINE WITH EARLY TREATMENT

The available cholinergic drugs not only improve memory and other cognitive functions but also benefit overall patient function and help manage some of the behavioral disturbances associated with dementia. Studies show that cholinergic drugs appear to have their greatest therapeutic effect in patients with mild to moderate disease, so early diagnosis is critical in order to help patients maintain the highest level of functioning they have left.

Dr. Murray Raskind, University of Washington, and his collaborators from other U.S. institutions studied what happens when drug treatment is delayed in Alzheimer's patients. They treated half of their volunteers with the cholinesterase inhibitor galantamine (Reminyl), and the other half took a sugar pill placebo. Six months later, the researchers began giving all the patients the active drug. The group of patients who had been on placebo showed rapid improvement, but they never tested as well as the patients with a six-month jump-start on the medication. In fact, the added benefit for the early starters appeared to continue for the entire year of follow-up, as illustrated in Figure A.3.

Figure A.3

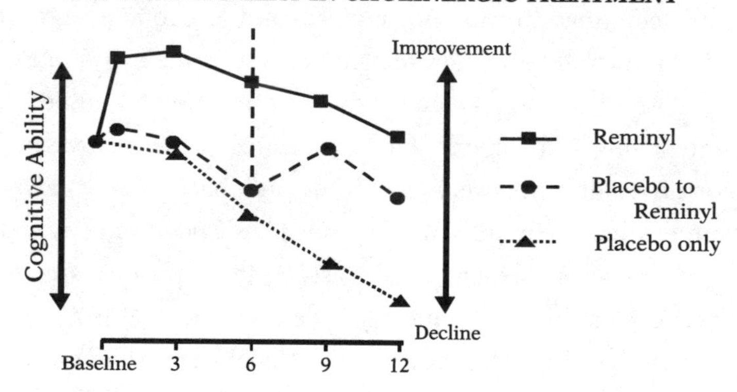

Cholinergic drugs enhance the body's level of acetylcholine, the chemical neurotransmitter that facilitates passage of nerve impulses across synapses. The brains of Alzheimer's patients have a deficiency of acetylcholine, which can result from either impaired production or excess breakdown by enzymes called cholinesterases, and our currently approved treatments inhibit these enzymes, so they are called cholinesterase inhibitors. Tacrine (Cognex) was the first of these medicines to be approved, but it is now rarely used due to the extent of its side effects. The newer compounds have fewer side effects and include donepezil (Aricept, approved by the FDA in November 1996), rivastigmine (Exelon, approved in April 2000), and galantamine (Reminyl, approved May 2001). These drugs not only improve memory and thinking but they can also reduce agitation and depression. Very recently, investigators have been studying their effects on different forms of dementia. Reminyl has been shown to be effective in patients with vascular dementia, and Exelon in patients with Lewy body dementia.

Although the majority of patients tolerate these drugs perfectly well, some have reported mild side effects including loss of appetite, indigestion, nausea, slowed heart rate, and insomnia. Most doctors increase the medications gradually in order to minimize side effects, which can occur when the medication is initially begun or the dosage increased. Side effects usually subside with time.

DOSING OF COMMONLY USED CHOLINESTERASE INHIBITORS

DRUG	START DOSE	HIGHEST DOSE
Donepezil (Aricept)	5 mg, once a day	10 mg, once a day
Galantamine (Reminyl)	4 mg, twice a day	12 mg, twice a day
Rivastigmine (Exelon)	1.5 mg, twice a day	6 mg, twice a day

When beginning cholinergic drug treatment, most patients show improvement, but after months or even a year, they eventually plateau and begin to gradually decline. The big mistake is to assume the drug is no longer working and discontinue treatment once the patient starts this gradual but inevitable decline. But not so fast—the patient *does* have Alzheimer's disease, remember? The drug is meant to treat symptoms and slow further decline. It is *not* a cure. Imagine this patient's decline over a twelve-month period without the drug. It would have been much more rapid. So if the patient is tolerating the medicine, stick with it. I advise patients to stay on their medication as long as they are tolerating it, since the overwhelming evidence shows that cholinesterase inhibitors can slow the rate of cognitive and functional decline even if patients don't experience obvious initial improvement on the drug.

Alzheimer's patients who take cholinergic drugs need fewer medications for treating depression and behavior problems. They also remain at home and out of nursing homes longer than patients who do not take these medicines.

VITAMIN E FOR EVERYBODY

If you have been a dutiful follower of the strategies in Chapter 7, you have been taking your antioxidant vitamins, including vitamin E, and doing your part to help your body fight those pesky free radicals that wear and tear down your body's DNA. Once Alzheimer's attacks the brain, vitamin E is as important as ever.

Recognizing the potential antioxidant benefits of vitamin E, Dr. Mary Sano, Columbia University, and her associates found that Alzheimer's patients showed less rapid functional decline if they took 1,000 international units (IUs) of vitamin E twice daily when compared to patients on placebo. Functional decline was defined as amount of time—days, weeks, months—before the patient needed nursing home care, as well as other practical indicators of daily function.

The scientists chose a very high dose of vitamin E to ensure that enough was present to have an effect. Because such high doses could occasionally suppress immune function and the patient's ability to ward off some infections, not all physicians recommend such high doses, preferring 800 to 1,000 IUs per day for patients suffering from Alzheimer's disease. As I mentioned in Chapter 7, I recommend 400 IUs once or twice daily for everyone as a preventative measure.

TREATING BEHAVIOR DISTURBANCES

Behavior changes in patients with advanced Alzheimer's disease can literally drive family members and caregivers into depression and their own health problems. If you are caring for someone who constantly yells at you or strikes out or watches you suspiciously all day, it is hard not to react with anger, guilt, or sadness. Unfortunately, caregivers often take the patient's behaviors personally and interpret them as willful rather than resulting from misfiring neurons. These kinds of behavioral changes are what usually lead to the placement of Alzheimer's patients into nursing homes. Caregivers just can't take it anymore.

Sometimes medications will improve symptoms, and non-medicinal approaches can make a big difference. When counseling caregivers on dealing with some of these problems, I often remind them of how we deal with young children. Similar strategies are useful in both situations: simplify communications, distract them when they get frustrated, and maintain a calm but firm attitude.

Antipsychotic drugs are often used to treat psychotic symptoms and agitation in patients with dementia. Newly developed antipsychotics are preferred, including risperidone (Risperdol), quetiapine (Seroquel), ziprasidone (Geodon), and olanzapine (Zyprexa). These newer drugs cause fewer side effects than older antipsychotics like haloperidol (Haldol) or chlorpromazine (Thorazine). For symptoms of anxiety and agitation, the anti-anxiety drugs are often used. Most physicians prefer the newer drugs, including alprazolam (Xanax), lorazepam (Ativan), and oxazepam (Serax) because older drugs like diazepam (Valium) and chlordiazepoxide (Librium) tend to accumulate in the blood and cause side effects like daytime sedation, unsteady gait, and confusion. The newer drugs get in and out of the body more quickly and tend to cause fewer side effects.

Some medicines used to treat epilepsy, the anticonvulsants—particularly carbamazepine (Tegretol) and divalproex sodium (Depakote)—represent another group of drugs that show promise as effective treatments for behavioral problems in demented patients. Dr. Pierre Tariot at the University of Rochester has led the field in using these drugs in this patient population and finds that agitated and aggressive patients who have a manic appearance of rapid thinking and irritability are especially responsive to anticonvulsant medications. For demented patients who develop depression, antidepressants are often prescribed (Chapter 9).

MEETING THE CHALLENGES WHEN THINGS GET TOUGH

If you are a caregiver, you may want to consider joining a support group to help answer your questions, make you feel less alone, and diminish your level of stress. Feelings of anger, frustration, and guilt are a normal part of caring for a relative or friend or patient with dementia. Community resources can offer some respite care, giving relatives and friends a chance to care for themselves, go to the gym, join a support group, see friends, and other personal activities to recharge their batteries.

Dr. Mary Mittelman and her colleagues at New York University have studied how education and emotional support for caregivers of Alzheimer's patients may delay the patient's placement into nursing homes. They found that the caregivers' education and support had a definite impact on their patients, delaying nursing home placement up to a year.

Establishing a daily routine for patients will improve their behavior and mood (see box, page 264). The predictability gives them a sense of security. Clocks and calendars help keep patients oriented. Consider setting up an exercise program that allows patients to move about freely for as long as possible. Newspapers, radios, and televisions are great ways for patients to try to stay up on current events and keep links to the outside world. Try to help them maintain social and intellectual activities and continue to attend family events whenever possible.

If behavior becomes troublesome, try to understand what provokes it. Some patients' symptoms get worse toward the evening, when rooms tend to darken. Other times a particular family member, friend, caregiver, or situation brings on aggressive outbursts. When possible, modify a situation to avoid provocation. Unfortunately, many patients become agitated out of mere frustration, or conversations become too complex, or perhaps they forget the content of the discussion. Try to use simple sentence structure and give reassurance by gently reminding the patient of the content of the discussion. If caregivers can stay calm, patients often pick up on their composure, which can help them to calm down as well.

For patients who tend to wander, using night-lights and perhaps even raising doorknobs up high may help to keep them safe. Regular supervised walks will promote exercise and may cut down on wandering. However, if a large yard is available so the patient has the space to walk about safely, medications or other forms of restraint can be reduced or avoided. Also contact the Alzheimer's Association (800-272-3900) for information and to register with the Safe Return program, which provides patient name tags and medical-alert bracelets that can help to locate lost patients.

The cognitive impairments of Alzheimer's disease diminish driving skills, and even mildly demented patients often should not be driving due to their difficulties with visual and spatial skills and their diminished capacity to plan ahead. Some states, such as California, require the physician to report patients with Alzheimer's disease to better monitor their driving skills. A diagnosis of dementia should clearly raise concern about a person's driving abilities, and patients with advanced dementia should not be driving at all.

The overall goal of caregiving is to maintain the kindest, least restrictive environment as possible, for as long as possible. For many families, this means keeping patients at home and out of long-term care facilities unless it becomes absolutely necessary.

What to Do If Alzheimer's Disease Strikes

- Seek professional help in attaining an accurate diagnosis so appropriate treatment can begin sooner rather than later.
- Ask the doctor about medication treatments, including cholinergic drugs and vitamin E. Discuss other practical and legal issues with the doctor.
- Join a support group for family members and caregivers through the Alzheimer's Association or local community groups where available.
- Expect that the patient's behavior will change over time and may become difficult. Remember that these changes are not willful but are a result of a physical brain disease.
- Maintain the patient's social and family activities as much as and as long as possible.
- Individualize sensory input to the patient's needs.
- Try to understand the cause of a patient's troublesome behavior and attempt to avoid it.
- Keep daily activities routine and surroundings familiar.
- Arrange a regular exercise schedule to promote fitness and minimize wandering. If necessary and possible, provide an environment where the patient can wander safely.
- Display clocks and calendars to help orient the patient.
- For the safety of the patient as well as everyone on the road, patients with moderate to severe dementia should not drive under any circumstances.

APPENDIX 3

Current and Potential Treatments for Memory Loss and Alzheimer's Disease

A wide range of drugs, hormones, herbs, and even surgical treatments have been proposed and tested for Alzheimer's disease and mild forms of age-related memory loss. For many of these treatments, the strongest evidence available for their effectiveness is based on testimonials, so they may not be any more effective than placebos. Thus, consumers need to proceed with caution when considering any new intervention. At best, they may be wasting their time and money. At worst, they may be exposing themselves to a risk of harmful side effects and drug interactions.

Although cholinesterase inhibitors are the only approved treatments for Alzheimer's disease, most experts also recommend antioxidant vitamins, and several promising approaches are currently in development. It is still important to keep an open mind about some of the alternative treatments—just because scientists have not yet proven them to be effective doesn't mean that they don't actually work.

Acetyl-l-carnitine. Acetyl-l-carnitine promotes the neurotransmitter acetylcholine and may protect brain neurons. Some human studies show a benefit over placebo for memory performance.

Ampakines. Drugs that increase the activity of brain chemicals important to memory formation. Ampakines are being tested in patients with Alzheimer's disease and mild cognitive impairment.

Anti-amyloid vaccine. A synthetic form of amyloid-beta used to vaccinate mice that have been genetically engineered to form the Alzheimer's plaques. The vaccine is currently being tested in humans to determine if it can treat or prevent Alzheimer's disease.

Cholinesterase inhibitors. Drugs approved by the FDA for the treatment of Alzheimer's disease. These drugs inhibit the cholinesterase enzymes, which break down acetylcholine, resulting in an increase in acetylcholine and improved cognition in patients. Current investigations are focusing on their potential for treating milder forms of memory loss and delaying the onset of Alzheimer's disease.

Clioquinoline. This antibiotic was once used to treat traveler's diarrhea and is now under investigation as a possible anti-amyloid treatment for Alzheimer's disease. Tests in transgenic Alzheimer's mice indicate clioquinoline attaches to the metals in the brain plaques and clears them out, leading to more than a 50 percent reduction in plaques, as well as improved general behavior. Scientists have begun comparing the drug with placebo in patients with Alzheimer's disease.

Cox II inhibitors. A new group of anti-inflammatory drugs that inhibit only one of the enzymes (cyclooxygenase II) involved in inflammation. These drugs have fewer side effects than older cyclooxygenase inhibitors, which inhibit both cox I and cox II enzymes and are more likely to cause gastric bleeding.

These drugs are currently being studied as treatments to prevent Alzheimer's disease.

DHEA (dehydroepiandrosterone). The body converts DHEA into estrogen and testosterone. DHEA supplements may strengthen the immune system and heighten sex drive and activity level. Side effects include an increased risk for prostate cancer, facial hair growth, scalp balding, and acne. Studies of its potential memory benefits are inconclusive.

Donepezil (Aricept). A cholinesterase inhibitor drug used to treat Alzheimer's disease. Donepezil should be increased from 5 to 10 mg daily after six weeks. Dose increases should be slower if patients have difficulties tolerating side effects.

Estrogen. Current studies will determine if estrogen replacement therapy after menopause lowers the risk for Alzheimer's disease. Studies of estrogen as a treatment for Alzheimer's disease have been disappointing, and it is not recommended for treating memory loss.

Fetal cell implants. A technology under development for growing new nerve cells in the brain. Results to date have not been successful.

Ginkgo biloba. A Chinese herb used to treat memory loss. Previous studies suggesting a benefit have not yet been confirmed.

Ginseng. An herb that has been considered as a treatment for memory loss because of its potential to enhance mental arousal. No systematic study has substantiated memory benefits.

Galantamine (Reminyl). A cholinesterase inhibitor drug used to treat Alzheimer's disease. Galantamine treatment should begin at 4 mg twice daily and increased every month for a maximum dose of 12 mg twice daily. Dose increases should be slower if patients have difficulties tolerating side effects.

Hydergine. Derived from a rye fungus, hydergine acts on several neurotransmitters that influence memory. It has been used extensively as a cognitive enhancer throughout the world. Studies of patients with dementia or with age-associated cognitive symptoms have yielded mixed results.

Lecithin. A major component of all living cells, lecithin is broken down in the body to active cholinergic compounds that have minimal but inconsistent memory benefits. The average diet contains about a gram of lecithin, and tenfold greater amounts are used as supplements given in daily divided doses.

Melatonin. The principal hormone secreted by the pineal gland, melatonin is a brain messenger with a structure similar to serotonin. The hormone regulates mood, sleep, sexual behavior, reproductive alterations, immunologic function, and the sleep-wake cycle. Its antioxidant activity has led investigators to consider it as a treatment against Alzheimer's disease.

Memantine. A drug that acts on the NMDA (N-methyl-D-asparate) brain receptor involved in memory function. The number of these receptors decreases in Alzheimer's disease, and recent studies indicate memantine's benefits in severely demented patients.

Nerve growth factor. A group of chemicals secreted by genetically engineered nerve cells, which stimulate neuron growth and boost the brain's cholinergic neurotransmitter system. A limitation of human use is the difficulty of getting nerve growth factor into the brain.

Nicotine patches. Transdermal nicotine applied through skin patches shows promise as a therapy for conditions ranging from age-associated memory impairment to Alzheimer's disease. Preliminary studies indicate that nicotine can improve short-term memory and attention in patients with Alzheimer's disease. No study has yet proven its long-term effectiveness.

Nimodipine. A drug that blocks cellular channels that transport the ion calcium, which can destroy brain cells. Nimodipine is marketed as a cognition-enhancing agent in Europe and is under investigation in the United States.

Nonsteroidal anti-inflammatory drugs (NSAIDs). Drugs that interfere with the body's inflammatory process, generally used to treat minor injuries and arthritis. Examples include aspirin, ibuprofen (Motrin, Advil, Feldene), and cox-II inhibitors (Celebrex, Vioxx).

Nootropics. A class of drugs, including piracetam, oxiracetam, pramiracetam, and aniracetam, that enhance brain circulation. An anti-dementia effect has not been established, and controlled studies have yielded mixed results.

Omental transposition. A surgical procedure wherein a membrane that surrounds the stomach and other abdominal organs is dragged up to the head and attached to the brain's surface. Its potential benefit is thought to result from stimulation of small blood vessel growth. Although anecdotal reports are encouraging, no systematic study has demonstrated its effectiveness in Alzheimer's disease.

Phosphatidylserine. A nutrient present in fish, green leafy vegetables, soy products, and rice, which some experts recommend as a supplement for age-associated memory impairment. Results of placebo-controlled studies have been positive but long-term benefits are not known. The recommended starting dose ranges from 200 to 300 mg daily followed by a maintenance dose of 100 mg daily after several months.

Physostigmine. A cholinesterase inhibitor drug with a very brief duration of action, such that the pills need to be taken every few hours for a memory effect. Although some studies demonstrate a mild benefit, it is not generally used and long-term effects are not known.

Rivastigmine (Exelon). A cholinesterase inhibitor drug used to treat Alzheimer's disease. Rivastigmine can be increased every two weeks beginning at 1.5 mg twice daily up to 6 mg twice a day or the highest dose tolerated. Dose increases should be slower if patients have difficulties tolerating side effects.

Secretase inhibitors. Drugs under development to counteract enzymes (secretases) that form the toxic amyloid-beta fragment.

Selective estrogen receptor modulators (SERMs). New synthetic estrogens designed to isolate beneficial hormone effects and eliminate side effects. Studies have not yet shown SERMs to benefit cognitive function, but many experts remain optimistic of their eventual utility.

Selegiline (Eldepryl). A drug that inhibits monoamine oxidase enzymes that destroy neurotransmitters and has an antioxidant effect. Selegiline has been found to delay functional decline in patients with Alzheimer's disease. Because this effect is similar to that of vitamin E—and selegiline is more expensive and has more side effects—vitamin E is the preferred antioxidant treatment for Alzheimer's disease.

Tacrine (Cognex). The first cholinesterase inhibitor approved by the FDA. The drug is rarely used today because of liver side effects and frequent dosing schedule.

Testosterone. In men with low testosterone levels, the hormone may improve memory performance, and it is currently under investigation as a treatment for cognitive impairment.

Vitamin C. An antioxidant vitamin that may offer protection against age-related cognitive decline. Many experts recommend 500 to 1,000 mg daily to slow age-related cognitive decline.

Vitamin E. An antioxidant vitamin prescribed in high doses (1,000 to 2,000 units daily) for patients with Alzheimer's disease. Epidemiological studies suggest that taking vitamin E supplements may slow age-related cognitive decline, and many experts recommend from 400 to 800 units daily as a preventative therapy.

APPENDIX 4

GLOSSARY

Acetylcholine. A neurotransmitter involved in memory, learning, and concentration. The cholinergic neurons that produce brain acetylcholine decline in normal aging and in Alzheimer's disease.

Active observation. The process of focusing attention so that new information is stored into memory.

Aerobic exercise. Exercise that gets hearts pumping faster and lungs breathing deeper so more oxygen is delivered to the body's cells. Examples of aerobic exercise include calisthenics, rapid walking, jogging, and swimming. Research suggests that aerobic conditioning benefits brain function in the frontal lobe.

Age-associated memory impairment. The term for the common memory changes that accompany normal aging, defined as a memory decline demonstrated by at least one standard memory test, along with a subjective awareness of memory changes.

Age-related cognitive decline. A condition of noticeable decline in mental ability without the presence of disease.

Alzheimer's disease. The most common form of dementia. Its onset is gradual and its course progressive. The physician can make a "probable" diagnosis, but a "definite" diagnosis is made only through autopsy or biopsy.

Amyloid-beta. A small molecule consisting of about forty amino acids strung together like a beaded necklace. Amyloid-beta is the building block of the insoluble protein that forms the core of Alzheimer's plaques, thought to be toxic to the brain.

Amyloid plaques. Collections of decayed material resulting from brain cell death and degeneration, present in high concentrations in the areas involved in memory in the Alzheimer's brain. The central area of a plaque contains insoluble collections of amyloid-beta protein.

Antioxidants. Drugs, vitamins, or foods that interfere with oxidative stress.

Apolipoprotein E (APOE). A gene on chromosome 19 that comes in three different forms (2, 3, and 4). One copy of the APOE-4 gene increases the risk for Alzheimer's disease and lowers the average age when people first develop symptoms. Two copies have the same effect, but more so.

Catechin. A potent antioxidant found in green teas.

Cerebral cortex. A thick folded layer of nerve cells that covers the cerebrum. The cortex is divided into parietal, temporal, frontal, and occipital lobes, and its cells are involved in learning, recall, and language function.

Cerebrum. The main portion of the brain considered the seat of conscious mental processes.

Computerized tomography (CT) scan. A computer-enhanced X ray that provides pictures of brain structure that can assist in the diagnosis of brain tumors, strokes, and blood clots.

Cognition (cognitive function). Mental function involving memory, language abilities, visual and spatial skills, intelligence, and reasoning.

Cognitive stress test. An experimental method to tease out subtle brain abnormalities. Volunteers perform memory tasks while a functional MRI scanner measures brain activity response to memory performance.

Coronary heart disease. Heart disease caused by buildup of plaque in the coronary blood vessels, which provide blood and nutrients to the heart muscles.

Cortisol. A stress hormone secreted by the adrenal glands. Chronically high levels of cortisol can impair memory performance.

Dementia. Impairment in memory and at least one other cognitive function (e.g., language, visual-spatial skill) to the extent that it interferes with daily life.

Dendrites. Short, branching extensions of neurons that receive impulses from other neurons when neurotransmitters stimulate them.

Endorphins. Hormones responsible for the mild euphoria we feel after aerobic exercise, often described as the body's own internal circulating antidepressant.

Enzyme. A protein that controls chemical reactions in the body.

Ephedra. An herb used as a stimulant or appetite suppressant, which may cause rapid heart rate and anxiety when combined with other stimulants like coffee.

Epidemiological studies. Studies of large numbers of participants that count up rates of disease and factors that might influence disease risk.

Executive control. Cognitive abilities such as planning, scheduling, coordination, and actively inhibiting information, generally mediated in the frontal and pre-frontal brain regions.

Free radicals. Ubiquitous molecules, also known as oxidants, present in the air we breathe, the food we eat, and the water we drink. Free radicals cause oxidative stress and wear down the genetic material or DNA of our cells. This process accelerates aging and contributes to chronic diseases like cancer and Alzheimer's.

Frontal lobe. The front part of the brain that mediates executive control.

Genes. The blueprint for life contained in all the body's cells, inherited from parents, and consisting of deoxyribonucleic acid (DNA). The molecular configuration of the double-helical strands of DNA is an alphabet key, a genotype, that programs our phenotype—who we are mentally and physically. A minute molecular change can have a dramatic effect on a person's risk for a particular disease.

Glucose. A simple sugar that is the main source of energy in the body's cells and results from the breakdown of foods we eat.

Glycemic index. A measure of how rapidly a food causes blood sugar levels to rise. This index ranks foods from 0 to 100, indicating whether the food raises blood sugar levels gradually (low scores) or rapidly (high scores).

Gram (gm). A measurement used for drugs and equivalent to one-twenty-eighth of an ounce.

Gray matter. The outer part of the brain that contains cell bodies.

Hippocampus. A seahorse-shaped brain structure involved in memory and learning, located in the temporal lobe of the brain (near the temples).

Hormones. Chemical messengers produced by glands and organs in the body and absorbed into the blood stream.

Hypercholesterolemia. Elevated blood levels of cholesterol. A risk factor for diseases affecting blood vessels in the heart, brain, and other body organs.

Hypertension (high blood pressure). A chronic disease that increases risk for circulatory problems, heart disease, and vascular dementia.

Immediate memory. Fleeting memories for sights, sounds, and other stimuli that last for milliseconds before moving into short-term memory.

Immune response. The body's mechanism for developing a memory for foreign or threatening materials or organisms. Specialized immune cells are primed initially so they will quickly produce antibodies to ward off infection. The immune system recognizes and destroys proteins like insoluble amyloid-beta that are not normally present.

Inflammation. The body's natural response to infection or stress, consisting of a mobilization of specialized cells to eliminate the offending foreign body.

Insulin. A hormone produced by the pancreas that gets sugar into cells.

Insulin resistance. The inability of cells to respond to insulin, resulting in high blood sugar levels.

Ischemia. Lack of oxygen to body tissue. In the brain, it can have several effects: if brief, it leads to transient ischemic attacks (TIAs) when the patient has a temporary loss of cognitive or motor function. When prolonged, it leads to death of brain cells and permanent deficits, known as strokes or cerebral vascular accidents (CVAs).

Kava kava. An herb used to reduce anxiety and stress. When combined with alcohol, it can augment intoxication.

Lesion. Any damage to body tissues or cells.

Linking. A memory technique that associates or connects two or more bits of information.

Long-term memory. Relatively permanent memories that have been organized and rehearsed.

LOOK, SNAP, CONNECT. A basic three-step memory technique that includes: (1) actively observing what you want to learn (*LOOK*), (2) creating mental snapshots of memories (*SNAP*), and (3) linking mental snapshots together (*CONNECT*).

Lycopene. A potent antioxidant present in high concentrations in tomatoes.

MacArthur Study of Successful Aging. A long-term study of aging that addressed positive rather than negative outcomes. Successful aging involves avoiding diseases, remaining engaged in life, and maintaining high physical and mental functioning.

Magnetic resonance imaging (MRI) scan. A brain-scanning technique that provides more detailed information on brain structure than CT scanning and that can be useful in diagnosing brain tumors, strokes, and blood clots. When modified, it can produce information on brain function, a technique known as functional MRI.

Major depression. A serious form of depression that can interfere with memory ability.

Mild cognitive impairment. A memory impairment similar to that observed in mild Alzheimer's disease, but not great enough to interfere with a person's ability to live independently. People with this condition have about a 15 percent chance of developing Alzheimer's disease each year.

Microglia cells. Cells that serve to clean up debris in the brain and mediate the immune response that causes brain inflammation.

Milligram (mg). A measurement used for drugs—one-one-thousandth of a gram.

Mozart Effect. The term used to describe the observation that passive listening to classical music benefits mental abilities.

Neurofibrillary tangles. Collections of decayed material resulting from brain cell death and degeneration, present in high concentrations in the areas involved in memory in the Alzheimer's brain.

Neurogenesis. New nerve-cell growth.

Neuron. Nerve or brain cell.

Neurotransmitter. A small molecule that serves as a brain messenger, allowing one neuron to communicate with another.

Nicotinic receptors. A small receptacle in nerve cells where brain messengers like nicotine or acetylcholine attach and communicate their information through a series of chemical reactions.

Neuropsychological tests. Standardized tests that measure memory, attention, and other cognitive abilities.

Nutraceuticals. Natural substances not regulated by the FDA and used as supplements, often to counteract the aging process.

Objective memory ability. An individual's memory ability as measured by standardized memory tests or neuropsychological tests

Omega-3 fatty acids. So-called "good fats" that keep brain cell membranes soft and flexible and come from fruits, leafy vegetables, nuts, fish, and supplements.

Omega-6 fatty acids. So-called "bad fats" that tend to make brain cell membranes more rigid and come from animal meat, whole milk, cheese, margarine, mayonnaise, processed foods, fried foods, and vegetable oils.

ORAC *(oxygen radical absorbency capacity).* The unit for a laboratory measuring technique that determines the ability of different foods to counteract oxidative stress. Foods with high ORAC scores may protect brain cells from the damage of oxygen radicals, or free radicals.

Oxidative stress. The wear and tear that free radicals cause to the body's cells through a chemical reaction in which oxygen reacts with another substance to cause a chemical transformation. Antioxidants counteract this process.

Parietal lobe. The area of the brain above and behind the temporal region (near the temples).

Pharmacogenetics. The emerging field of drug effectiveness and safety based upon an individual's genetic makeup.

Positron emission tomography (PET). A body-scanning method that measures structure as well as function. PET scans of the brain show characteristic patterns of decreased metabolism in the areas affected by Alzheimer's disease.

Protein. A molecule that is the building block of neurotransmitters and enzymes.

Serotonin. A neurotransmitter necessary for relaxation, concentration, and sleep that is decreased in depression and dementia.

Short-term memory. Memories lasting only minutes and too transient for long-term recall.

Smart drugs. Medications, herbs, hormones, or supplements taken with the intent of improving memory and other cognitive functions in a normal individual who does not have obvious memory loss.

Statins. Cholesterol-lowering drugs that may reduce the risk for developing Alzheimer's disease.

St. John's wort. An herbal treatment for anxiety, depression, and insomnia.

Stress response. The body's physiological reaction to stress, involving release of cortisol and other stress hormones.

Stroke. Death of brain cells resulting in a loss of physical or mental function or both.

Subdural hematomas. Blood clots surrounding the brain; a potential side effect of ginkgo biloba.

Subjective memory ability. A person's self-awareness of memory ability.

Synapse. The interface between two nerve cells where they communicate information.

Valerian. An herb used for restlessness and insomnia, which can interact adversely with alcohol and other sedatives.

Vascular dementia. A dementia resulting from many small strokes.

Verbal memory. Learning and recall of information relating to language and words.

Visual-spatial memory. Learning and recall of visual and spatial information.

White matter. Brain areas consisting of nerve fibers that transfer information from distant brain regions.

APPENDIX 5

Additional Resources

Many organizations provide information on memory and general health issues important to maintaining brain health. Several national organizations also have local or state chapters. Check your telephone directory or Internet search engine for related organizations and websites.

Name & Address	Description	Telephone
AARP 6601 E Street NW Washington, DC 20049 *www.aarp.org*	Non-profit, non-partisan organization dedicated to helping older Americans achieve lives of independence, dignity, and purpose.	202-434-2277 800-424-3410

Name & Address	Description	Telephone
Academy of Molecular Imaging Box 951735 Los Angeles, CA 90095-1735 *www.ami-imaging.org*	Provides leadership in research and clinical aspects of molecular imaging of the biological nature of disease. Their website includes a listing of local PET centers.	310-267-2614
Administration on Aging 330 Independence Avenue NW Washington, DC 20201 *www.aoa.dhhs.gov*	Provides information for older Americans and their families on opportunities and services to enrich their lives and support their independence.	202-619-7501 800-677-1116
Aging Network Services 4400 E. West Highway, Suite 907 Bethesda, MD 20814 *ww.agingnets.com*	Nationwide network of private-practice geriatric social workers serving as care managers for seniors living at a distance.	301-657-4329
Alliance of Information and Referral Systems P.O. Box 31668 Seattle, WA 98103 *www.airs.org*	Professional organization that provides human services information and referrals.	206-632-2477

Name & Address	Description	Telephone
Alzheimer Europe 145 Route de Thionville L-2611 Luxembourg *www.alzheimer-europe.org*	Organizes caregiver support and raises awareness about dementia through cooperation among European Alzheimer organizations.	352-29-79-70
Alzheimer's Association 919 N. Michigan Ave., Suite 1000 Chicago, IL 60611-1676 *www.alz.org*	The national organization that provides information on services, programs, publications, and local chapters.	800-272-3900
Alzheimer's Association Public Policy Division 1319 F St. NW, Suite 710 Washington, D.C. 20004 *www.alz.org*	Lobbying branch of the Alzheimer's Association.	202-393-7737
Alzheimer's Disease Education and Referral Center P.O. Box 8250 Silver Spring, MD 20907 *www.alzheimers.org*	National Institute on Aging service that distributes information and free materials on topics relevant to health professionals, patients and their families, and the general public.	301-495-3311 800-438-4380

Name & Address	Description	Telephone
Alzheimer Research Forum Foundation 82 Devonshire Street, S3 Boston, MA 02109 *www.alzforum.org*	Provides information and promotes collaboration among researchers in order to foster a global effort to understand and treat Alzheimer's disease.	
American Academy of Neurology 1080 Montreal Avenue St. Paul, MN 55116 *www.aan.com*	Professional organization that advances the art and science of neurology, thereby promoting the best possible care for patients with neurological disorders.	651-695-1940
American Association for Geriatric Psychiatry 7910 Woodmont Ave. #1050 Bethesda, MD 20814 *www.aagpgpa.org*	Professional organization dedicated to enhancing the mental health and well-being of older adults through education and research.	301-654-7850
American Diabetes Association P.O. Box 25757 1660 Duke Street Alexandria, VA 22314 *www.diabetes.org*	America's leading non-profit health organization providing diabetes research, information, and advocacy.	703-549-1500 800-232-3472

Name & Address	Description	Telephone
American Dietetic Association 216 W. Jackson Blvd. Chicago, IL 60606-6995 *www.eartright.org*	Consumer Nutrition Hotline that provides information on finding a dietitian.	312-899-0040 800-366-1655
American Geriatrics Society 770 Lexington Avenue #300 New York, NY 10021 *www.americangeriatrics.org*	Professional association providing assistance in identifying local geriatric physician referrals.	212-308-1414 800-247-4779
American Heart Association 7272 Greenville Avenue Dallas, TX 75231 *www.americanheart.org*	Non-profit health organization whose mission is to reduce disability and death from cardiovascular diseases and stroke.	214-373-6300
American Society on Aging 833 Market Street, Suite 511 San Francisco, CA 94103 *www.asaging.org*	National organization concerned with physical, emotional, social, economic, and spiritual aspects of aging.	415-974-9600 800-537-9728

Name & Address	Description	Telephone
Children of Aging Parents 1609 Woodbourne Rd., #302-A Levittown, PA 19057 *www.caps4caregivers.org*	National organization providing information and referrals for caregivers of older adults.	215-945-6900 800-227-7294
Family Caregiver Alliance 425 Bush Street, Suite 500 San Francisco, CA 94108 *www.caregiver.org*	Resource center for families of adults with brain damage or dementia, which provides publications for caregivers and professionals.	415-434-3388 800-445-8106
Gerontological Society of America 1030 15th Street NW, Suite 250 Washington, DC 20005 *www.geron.org*	National interdisciplinary organization on research and education in aging.	202-842-1275
National Institute of Mental Health 5600 Fishers Lane, Room 10-75 Rockville, MD 20857 *www.nimh.nih.gov*	Part of the National Institutes of Health, the NIMH is the principal biomedical and behavioral research agency of the United States government.	301-443-1185

Name & Address	Description	Telephone
National Institute of Neurological Disorders and Stroke Building 31, Room 8A-06 31 Center Drive, MSC 2540 Bethesda, MD 20892-2540 *www.ninds.nih.gov*	The National Institutes of Health agency that supports neuroscience research; focuses on rapidly translating scientific discoveries into prevention, treatment, and cures; and provides resource support and information.	301-496-5751 800-352-9424
National Institute on Aging Building 31, Room 5C27 31 Center Drive Bethesda, MD 20892-2292 *www.nih.gov/nia*	The National Institutes of Health agency that supports research on aging and provides information about national Alzheimer's centers, and a free directory of organizations that serve older adults.	301-496-1752 800-438-4380
National Stroke Association 96 Inverness Drive East, Suite 1 Englewood, CO 80112-5112 *www.stroke.org*	Their mission is to reduce the incidence and impact of stroke disease and improve quality of patient care and outcomes.	303-649-9299 800-787-6537

Name & Address	Description	Telephone
Older Women's League 666 11th St. NW, Suite 700 Washington, D.C. 20001 *www.owl-national.org*	An advocacy organization addressing family and caregiver issues.	202-783-6686 800-825-3695
Safe Return P.O. Box 9307 St. Louis, MO 63117-0307 *www.alz.org*	Joint program of the Alzheimer's Association and the National Center for Missing Persons that provides patients with a dementia bracelet showing the person's name, the registered caregiver's name, and a toll-free number (800-572-1122) to aid in that person's return if lost.	888-572-8566
SeniorNet 121 Second Street, 7th Floor San Francisco, CA 94105 *www.seniornet.com*	A national non-profit organization that works to build a community of computer-using seniors.	415-495-4990

Name & Address	Description	Telephone
UCLA Center on Aging 10945 Le Conte Avenue, #3119 Los Angeles, CA 90095-6980 *www.aging.ucla.edu*	University center that works to enhance and extend productive and healthy life through research and education on aging.	310-794-0676
U.S. Department of Veterans Affairs 1120 Vermont Avenue NW Washington, DC 20421 *www.va.gov*	Provides information on VA programs, veterans benefits, VA facilities worldwide, and VA medical automation software.	800-827-1000

Bibliography

Alexopoulos GS, Meyers BS, Young RC, et al. The course of geriatric depression with "reversible dementia": A controlled study. *Am J Psychiatry* 1993;150:1693–1699.

Amaducci L, and the SMID group. Phosphatidylserine in the treatment of Alzheimer's disease: Results of multicenter study. *Psychopharm Bull* 1988;24:922–931.

Ancelin M-L, De Roquefeuil G, Ledesert B, et al. Exposure to anaesthetic agents, cognitive functioning and depressive symptomatology in the elderly. *Br J Psychiatry* 2001;178:360–366.

Anthony GA, Grayson DA, Creasey HM, et al. Anti-inflammatory drugs protect against Alzheimer disease at low doses. *Arch Neurol* 2000;57:1586 1591.

Argyriou A, Prast H, Philippu A. Melatonin facilitates short-term memory. *Eur J Pharm* 1998; 349:159–162.

Atherton D. Towards the safer use of traditional remedies. *Br Med J* 1994;308:673–674.

Ayd FJ. Evaluating interactions between herbal and psychoactive medications. *Psychiatric Times* 2000;December:45–47.

Barone JJ, Roberts HR. Caffeine consumption. *Food Chemistry and Toxicology* 1996;34:119–129.

Benowitz NL. Pharmacology of caffeine. *Ann Rev Med* 1990; 41:277–288.

Benson H. *The Relaxation Response*. Avon, New York, 1975.

Beyer CE, Steketee JD, Saphier D. Antioxidant properties of melatonin—an emerging mystery. *Biochem Pharm* 1998;56:1265–1272.

Bookheimer SY, Strojwas MH, Cohen MS, et al. Brain activation in people at genetic risk for Alzheimer's disease. *N Engl J Med* 2000; 343:450–456.

Bowen JD, Larson EB. Drug-induced cognitive impairment: Defining the problem and finding solutions. *Drugs and Aging* 1993;3:349–357.

Braak H, Braak E. Neuropathological staging of Alzheimer-related changes. *Acta Neuropathol* 1991;82:239–259.

Brand-Miller J, Volwever TMS, Colaguiri S, Foster-Powell K. *The Glucose Revolution*. Marlow & Co., New York, 1999.

Caffarra P, Santamaria V. The effects of phosphatidylserine in patients with mild cognitive decline. *Clinical Trials Journal* 1987;24:109–114.

Carper J. *Your Miracle Brain*. Harper Collins, New York, 2000.

Collins MW, Grindel SH, Lovell MR, et al. Relationship between concussion and neuropsychological performance in college football players. *JAMA* 1999;282:964–970.

Corder EH, Saunders AM, Strittmatter WJ, et al. Gene dose of apolipoprotein E type 4 allele and the risk of Alzheimer's disease in late onset families. *Science* 1993;261:921–923.

Crook TH, Adderly B. *The Memory Cure*. Pocket Books, New York, 1998.

Del Ser T, Hachinski V, Merskey H, Munoz DG. An autopsy-verified study of the effect of education on degenerative dementia. *Brain* 1999;122:2309–2319.

Devanand DP. *The Memory Program*. John Wiley & Sons, New York, 2001.

Eisenberg DM, Kessler RC, Foster C, et al. Unconventional medicine in the United States: Prevalence, costs, and patterns of use. *N Engl J Med* 1993;328:246–252.

Etherton GM, Kochar MS. Coffee: Facts and controversies. *Arch Family Med* 1993;2:317–322.

Feltrow CW, Avila JR. *The Complete Guide to Herbal Medicines*. Pocket Books, New York, 2000.

Fitten LJ, Perryman KM, Wilkinson CJ, et al. Alzheimer and vascular dementias and driving: A prospective road and laboratory study. *JAMA* 1995; 273:1360–1365.

Garattini S (ed.): *Caffeine, Coffee, and Health*. Raven Press; New York, 1993.

Ghebremedhin E, Schultz C, Braak E, Braak H. High frequency of apolipoprotein E ε4 allele in young individuals with very mild Alzheimer's disease-related neurofibrillary changes. *Exp Neurol* 1998;153:152–155.

Gilewski MJ, Zelinski EM, Schaie KW. The Memory Functioning Questionnaire for assessment of memory complaints in adulthood and old age. *Psychology and Aging* 1990;5:482–490.

Gwyther LP, Rabins PV. Practical approaches for treating behavioral symptoms of people with mild to moderate Alzheimer's disease. *Primary Psychiatry* 1996; 3:27–38.

Haier RJ, Siegel BV, MacLachlan A, et al. Regional glucose metabolic changes after learning a complex visuospatial/motor task: A positron emission tomographic study. *Brain Research* 1992;570:134–143.

Hall W, Solowij N, Lemon J. Health and psychological consequences of cannabis use. *1994 National Drug Strategy Monograph Series No. 25*. Canberra, Australian Government Publication Service.

Hathcock JN. Vitamins and minerals: Efficacy and safety. *Am J Clin Nutrition* 1997;66:427–437.

Heber D. *The Resolution Diet*. Avery, New York, 1999.

Heber D, Bowerman S. *What Color Is Your Diet?* Regan Books, New York, 2001.

Hebert LE, Scherr PA, Beckett LA, et al. Age-specific incidence of Alzheimer's disease in a community population. *JAMA* 1995;273:1354–1359.

High KP. Micronutrient supplementation and immune function in the elderly. *CID* 1999;28:717–722.

Ivgi M, Schnaider B, Rabinowitz J, Davidson M. A naturalistic study comparing the efficacy of a memory enhancement course to a general academic course. *Int Psychogeriatr* 1999;11:281–287.

Janus C, Pearson J, McLaurin J, et al. A beta peptide immunization reduces behavioural impairment and plaques in a model of Alzheimer's disease. *Nature* 2000;408:979–982.

Jarvik LF, Small GW. *Parentcare*. Crown, New York, 1988.

Jick H, Zornberg GL, Jick SS, et al. Statins and the risk of dementia. *Lancet* 2000;356:1627–1631.

Johansson C, Skoog I. A population-based study on the association between dementia and hip fractures in 85-year-olds. *Aging* 1996;8: 189–196.

Joseph JA, Nadeau D, Underwood A. *The Color Code: A Revolutionary Eating Plan for Optimum Health*. Hyperion, New York, 2002.

Kahn RL, Rowe JW. *Successful Aging*. Pantheon, New York, 1998.

Kramer AF, Hahn S, McAuley E, et al. Exercise, aging and cognition: Healthy body, healthy mind? In Fisk AD, Rogers W (eds.), *Human Factors Interventions for the Health Care of Older Adults*. Erlbaum, Hillsdale, NJ, 2001.

Lim GP, Chu T, Yang F, et al. The curry spice curcumin reduces oxidative damage and amyloid pathology in an Alzheimer transgenic mouse. *J Neurosci* 2001;21:8370–8377.

Lorayne H. *How to Develop a Super Power Memory*. Lifetime Books, Hollywood, FL, 1998.

Marseille DM, Silverman DHS, Hussain SA, et al. Cerebral hypometabolism characteristic of Alzheimer's disease occurs early enough to affect ultimate educational level. *J Nucl Med* 2001;42:146P (No. 546).

Matser JT, Kessels AG, Jordan BD, et al. Chronic traumatic brain injury in professional soccer players. *Neurology* 1998;51:791–796.

Matser JT, Kessels AG, Lezak MD, et al. Neuropsychological impairment in amateur soccer players. *JAMA* 1999;282:971–973.

McCutcheon LE. Another failure to generalize the Mozart effect. *Psychological Reports* 2000;878:325–330.

McDaniel LD, Lukovits T, McDaniel KD. Alzheimer's disease: The problem of incorrect clinical diagnosis. *J Geriatr Psychiatry Neurol* 1993;6:230–234.

McGaugh JL. Memory—A century of consolidation. *Science* 2000; 287:248–251.

McGeer PL, Schulzer M, McGeer EG. Arthritis and anti-inflammatory agents as possible protective factors for Alzheimer's disease: A review of 17 epidemiologic studies. *Neurology* 1996;47:425–432.

McKeith I, Del Ser T, Spano P, et al. Efficacy of rivastigmine in dementia with Lewy bodies: A randomised, double-blind, placebo-controlled international study. *Lancet* 2000;356:2024–2025.

Merchant C, Tang MX, Albert S, et al. The influence of smoking on the risk of Alzheimer's disease. *Neurology* 1999;52:1408–1412.

Mittelman MS, Ferris SH, Shulman E, et al. A family intervention to delay nursing home placement of patients with Alzheimer's disease. A randomized, controlled trial. *JAMA* 1996;276:1725–1731.

Mohs RC, Ashman TA, Jantzen K, et al. A study of the efficacy of a comprehensive memory enhancement program in healthy elderly persons. *Psychiatry Research* 1998;77:183–195.

Morris MC, Beckett LA, Scherr PA, et al. Vitamin E and vitamin C supplement use and risk of incident Alzheimer disease. *Alzheim Dis Associated Disorders* 1998;12:121–126.

Natural Medicines Comprehensive Database. www.naturaldatabase.com.

The Natural Pharmacist. www.tnp.com.

Newman PE. Could diet be used to reduce the risk of Alzheimer's disease? *Medical Hypotheses* 1998;50:335–337.

Paivio A, Yuille JC, Madigan SA. Concreteness, imagery, and meaningfulness values for 925 nouns. *J Exp Psychol Monograph Suppl* 1968;76 (no.1, part 2):1–25.

Paykel ES, Brayne C, Huppert FA, et al. Incidence of dementia in a population older than 75 years in the United Kingdom. *Arch Gen Psychiatry* 1994;51:325–332.

Price JL, Morris JC. Tangles and plaques in nondemented aging and "preclinical" Alzheimer's disease. *Ann Neurol* 1999;45:358–368.

Raskind MA, Peskind ER, Wessel T, and the Galantamine USA-1 Study Group. Galantamine in AD: A 6-month randomized, placebo-controlled trial with a 6-month extension. *Neurology* 2000;54:2269–2276.

Rauscher FH, Shaw GL. Key components of the Mozart effect. *Perceptual and Motor Skills* 1998;86:835–841.

Reiman EM, Caselli RJ, Chen K, et al. Declining brain activity in cognitively normal apolipoprotein E ε4 heterozygotes: A foundation for using positron emission tomography to efficiently test treatments to prevent Alzheimer's disease. *Proc Natl Acad Sciences USA* 2001;98:3334–3339.

Reiman EM, Caselli RJ, Yun LS, et al. Preclinical evidence of Alzheimer's disease in persons homozygous for the epsilon 4 allele for apolipoprotein E. *N Engl J Med* 1996;334:752–758.

Relkin NR, Tanzi R, Breitner J, et al. Apolipoprotein E genotyping in Alzheimer's disease: Position statement of the National Institute on Aging/Alzheimer's Association Working Group. *Lancet* 1996;347:1091–1095.

Reneman L, Lavalaye J, Schmand B, et al. Cortical serotonin transporter density and verbal memory in individuals who stopped using 3,4-methylenedioxymethamphetamine (MDMA or "ecstasy"). *Arch Gen Psychiatry* 2001;58:901–906.

Rideout BE, Dougherty S, Wernert L. Effect of music on spatial performance: A test of generality. *Perceptual and Motor Skills* 1998;86: 512–524.

Ritchie K. Mental status examination of an exceptional case of longevity: J.C. aged 118 years. *Br J Psychiatry* 1995;116:229–235.

Ritchie K, Kildea D. Is senile dementia "age related" or "aging related"?: Evidence from a meta-analysis of dementia prevalence in the oldest old. *Lancet* 1995;346:931–934.

Rogers SL, Friedhof LT, Apter JT, et al. The efficacy and safety of donepezil in patients with Alzheimer's disease: Results of a US multicentre, randomized, double-blind, placebo-controlled trial. *Dementia* 1996;7:293–303.

Rondeau V, Commenges D, Jacqmin-Gadda H, Dartigues JF. Relation between aluminum concentrations in drinking water and Alzheimer's disease: An 8-year follow-up study. *Am J Epidem* 2000;152:59–66.

Roses AD, Saunders AM. ApoE, Alzheimer's disease, and recovery from brain stress. *Ann NY Acad Sci* 1997;826:200–212.

Sano M, Ernesto C, Thomas RG, et al. A controlled trial of selegiline, alpha-tocopherol, or both as treatment for Alzheimer's disease. *N Engl J Med* 1997;336:1216–1222.

Satoh T, Sakurai I, Mihagi K, Hohsaku Y. Walking exercise and improved neuropsychological functioning in elderly patients with cardiac disease. *J Internal Med* 1995;238:423–428.

Schenk D, Barbour R, Dunn W, et al. Immunization with amyloid-beta attenuates Alzheimer-disease-like pathology in the PDAPP mouse. *Nature* 1999;400;173–177.

Schneider LS, Tariot PN, Small GW. Update on treatment for Alzheimer's disease and other dementias. In Dunner DL, Rosenbaum JF (eds.), *Psychiatric Clinics of North America: Annual of Drug Therapy* 1997; 4:135–166.

Schofield PW, Marder K, Dooneief G, et al. Association of subjective memory complaints with subsequent cognitive decline in community-dwelling elderly individuals with baseline cognitive impairment. *Am J Psychiatry* 1991;154:609–615.

Schor JB. *The Overworked American*. Basic Books, New York, 1991.

Selye H. *Stress Without Distress*. Signet, Scarborough, 1975.

Shoghi-Jadid K, Small GW, Agdeppa ED, et al. Localization of neurofibrillary tangles (NFTs) and beta-amyloid plaques (APs) in the brains of living patients with Alzheimer's disease. *Am J Geriatr Psychiatry* 2002;10:24–35.

Silverman DHS, Ercoli LM, Huang S-C, et al. Perceived loss of memory function correlates with subsequent decline in metabolism of hippocampus. *J Nucl Med* 2001;42:61P (No. 228).

Silverman DHS, Small GW, Chang CY, et al. Evaluation of dementia with positron emission tomography: Regional brain metabolism and long-term outcome. *JAMA* 2001; 286: 2120–2127.

Small GW. Estrogen effects on the brain. *J Gender-Specific Med* 1998;1:23–27.

Small GW. Living better longer through technology. *Int Psychogeriatr* 1999;11:3–6.

Small GW. The pathogenesis of Alzheimer's disease. *J Clin Psychiatry* 1998;59(suppl 9):7–15.

Small GW. Positron emission tomography scanning for the early diagnosis of dementia. *West J Med* 1999;171:293–294

Small GW. Recognizing and treating anxiety in the elderly. *J Clin Psychiatry* 1997;58(suppl 3):41–47.

Small GW. Treatment of Alzheimer's disease: Current approaches and future prospects. In Dunner DL, Rosenbaum JF (eds.), *Psychiatric Clinics of North America: Annual of Drug Therapy* 2000;7:201–218.

Small GW, Donohue JA, Brooks RL. An economic evaluation of donepezil in the treatment of Alzheimer's disease. *Clin Therapeutics* 1998;20:838–850.

Small GW, Ercoli LM, Silverman DHS, et al. Cerebral metabolic and cognitive decline in persons at genetic risk for Alzheimer's disease. *Proc Natl Acad Sciences USA* 2000;97:6037–6042.

Small GW, Leuchter AF, Mandelkern MA, et al. Clinical, neuroimaging, and environmental risk differences in monozygotic female twins appearing discordant for dementia of the Alzheimer type. *Arch Neurol* 1993;50:209–219.

Small GW, Mazziotta JC, Collins MT, et al. Apolipoprotein E type 4 allele and cerebral glucose metabolism in relatives at risk for familial Alzheimer disease. *JAMA* 1995;273:942–947.

Small GW, Rabins PV, Barry PP, et al. Diagnosis and treatment of Alzheimer's disease and related disorders: Consensus statement of the American Association for Geriatric Psychiatry, the Alzheimer's Association, and the American Geriatrics Society. *JAMA* 1997;278:1363–1371.

Snowdon DA, Kemper SJ, Mortimer JA, et al. Linguistic ability in early life and cognitive function and Alzheimer's disease in late life: Findings from the Nun Study. *JAMA* 1996;275:528–532.

Specialty Coffee Association of American: Document #1040: Forecast for Continued Growth, in Lingle TR. *Avenues for Growth: A 20 Year Review of the U.S. Specialty Coffee Market.* SCAA, Long Beach, CA, 1993.

Spencer JW, Jacobbs JJ (eds). *Complementary/Alternative Medicine: An Evidence-Based Approach.* Mosby-Year Book, St. Louis, 1999.

Steele KM, Brown JD, Stoecker JA. Failure to confirm the Rauscher and Shaw description of recovery of the Mozart effect. *Perceptual and Motor Skills* 1999;88:843–848.

Swan GE, DeCarli C, Miller BL, et al. Association of midlife blood pressure to late-life cognitive decline and brain morphology. *Neurology* 1998;51:986–993.

Thompson GW. Coffee: Brew or bane? *Am J Med Sci* 1994;308:49–57.

Whalley LJ, Starr JM, Athawes R, et al. Childhood mental ability and dementia. *Neurology* 2000;55:1455–1459.

Winblad B, Poritis N. Memantine in severe dementia: Results of the 9M-Best study (Benefit and efficacy in severely demented patients during treatment with memantine). *Int J Geriatr Psychiatry* 1999;14:135–146.

World Health Organization International Agency for Research on Cancer, Coffee, Tea, Mate, Methylxanthines and Methylgloxal. *IARC Monogr Eval Carcinog Risks Hum.* 1991;51:1–513.

Zelinski EM, Gilewski MJ, Anthony-Bergstone CR: Memory Functioning Questionnaire: Concurrent validity with memory performance and self-reported memory failures. *Psychology and Aging* 1990;5:388–399.

Source Credits

Chapter 2:

Subjective Memory Questionnaire adapted with permission from the work of Dr. Michael Gilewski (Gilewski et al. 1990).

Chapters 3 and 6:

Drawings by Diana Jacobs

Chapter 5:

Mental aerobics exercises were included with permission from:

Marge Engelman and AgeNet.Inc (Aerobics of the Mind: *http://agenet.agenet.com*): All warm-up exercises, beginning exercise 2, and intermediate exercises 2, 3, and 1.

Kevin N. Stone (Brainbashers: *www.brainbashers.com*): Beginning exercise 7, intermediate exercise 9, and advanced exercises 4, 5, and 7.

The Ultimate Puzzle Site (*www.dse.nl/puzzle/index_us.html*): Beginning exercises 8 and 14.

The Grey Labyrinth (www.greylabyrinth.com): Advanced exercises 7 and 8.

Chapter 9

Figure 9.1 from Rogers SL, Friedhof LT, Apter JT, et al. The efficacy and safety of donepezil in patients with Alzheimer's disease: results of a US multicentre, randomized, double-blind, placebo-controlled trial. *Dementia* 1996;7:293–303.

Appendix:

Amyloid-PET scans (Figure A.1) courtesy of Jorge Barrio, Ph.D., Henry Huang, Ph.D., UCLA Department of Molecular and Medical Pharmacology.

PET scans showing brain function patterns (Figure A.2) courtesy of Michael Phelps, Ph.D.; Diane Martin, UCLA Department of Molecular and Medical Pharmacology.

Graph showing six-month delay in cholinergic treatment (Figure A.3) adapted from Raskind et al. 2000.

INDEX